Cloning

Recent Titles in the
CONTEMPORARY WORLD ISSUES
Series

Books in the **Contemporary World Issues** series address vital issues in today's society such as genetic engineering, pollution, and biodiversity. Written by professional writers, scholars, and nonacademic experts, these books are authoritative, clearly written, up-to-date, and objective. They provide a good starting point for research by high school and college students, scholars, and general readers as well as by legislators, businesspeople, activists, and others.

Each book, carefully organized and easy to use, contains an overview of the subject, a detailed chronology, biographical sketches, facts and data and/or documents and other primary source material, a forum of authoritative perspective essays, annotated lists of print and nonprint resources, and an index.

Readers of books in the Contemporary World Issues series will find the information they need to have a better understanding of the social, political, environmental, and economic issues facing the world today.

Cloning

A REFERENCE HANDBOOK

David E. Newton

ABC-CLIO™

An Imprint of ABC-CLIO, LLC
Santa Barbara, California • Denver, Colorado

Library of Congress Cataloging-in-Publication Data

Newton, David E.
    Cloning : a reference handbook / David E. Newton.
      pages cm. — (Contemporary world issues)
    Includes bibliographical references and index.
    ISBN 978-1-61069-693-7 (alk. paper) —
ISBN 978-1-61069-694-4 (ebook)   1. Cloning.   2. Human
cloning.   3. Clones (Plants)   4. Plant cuttings.   5. Pets—
Cloning.   6. Extinct animals—Cloning.   I. Title.
    QH442.2.N495 2015
    571.8—dc23   2015015707

ISBN: 978-1-61069-693-7
EISBN: 978-1-61069-694-4

19   18   17   16   15   1   2   3   4   5

This book is also available on the World Wide Web as an eBook.
Visit www.abc-clio.com for details.

ABC-CLIO
An Imprint of ABC-CLIO, LLC

ABC-CLIO, LLC
130 Cremona Drive, P.O. Box 1911
Santa Barbara, California 93116–1911

This book is printed on acid-free paper ∞

Manufactured in the United States of America

Cloning is one of the oldest technical procedures known to humankind. For at least two thousand years, men and women who have tilled the soil and grown their own crops have understood that it is possible to change the characteristic properties of plants by means of a variety of forms of asexual reproduction, reproductive technologies that do not require the combination of male and female gametes (sperm and eggs). Those technologies have flowed downward through human history to the present day, where farmers still use techniques such as grafting, layering, budding, rooting, and plant tissue culture to produce new types of plants that are more resistant to pests and diseases, that taste or look better than their ancestors, that are more nutritious to eat, that store longer on grocery shelves, or that have other desirable properties. Indeed, there is hardly a commercial food plant available to consumers today whose history has not included some type of improvement by way of cloning technology.

Cloning did not become an issue of purely scientific interest, however, until well into the 19th century when researchers began to ask more fundamental questions as to how asexual reproduction actually occurs, what parts of a plant can be used in cloning, and how humans can manipulate the environment in which asexual reproduction occurs most efficiently. And it was not until the dawn of the 20th century that researchers began to ask similar questions about the cloning of animals.

Even then, it was not until well after the middle of the century that animal cloning became anything other than almost purely speculation, a matter of interest to pure researchers with little or no known practical application in everyday life.

That situation began to change rapidly toward the end of the 20th century when researchers began to announce the cloning of the first fish (1963); the first mouse (1979); the first mammal (1986); the first cloned cattle, a cow named Gene (1997); the first cloned goat named Mira (1998); the first cloned pig, Xena (2000); the first cloned member of an endangered species, a gaur (2001); and the first cloned member of an extinct species, a Pyrenean ibex (2009). The possibility of cloning at least one more important species not mentioned in the list, a human, became at least theoretically possible in 1998 when researchers at the University of Wisconsin, Madison, and Johns Hopkins University first isolated human embryonic stem cells and kept them alive long enough to conduct research on them.

At this point in history, cloning stopped being a somewhat intriguing type of research that was of use primarily to agricultural and dairy scientists, with little impact on the lives of ordinary humans. News that researchers now knew how to make human embryonic stem cells, with the capability of differentiating into any and all types of human somatic cells, startled nearly everyone who read or heard about the development. Immediately, a host of scientific, technical, ethical, social, economic, psychological, and other questions were raised as to whether such research should even be permitted and, if so, under what circumstances.

Out of that debate grew the great separation between the use of cloning technology for the reproduction of human beings, an option opposed by virtually everyone around the world, and for therapeutic purposes, for the production of stem cells that could in theory be used to treat a host of otherwise intractable human diseases, such Alzheimer's disease, Parkinson's disease, a whole range of neurodegenerative diseases, and diabetes. Public opinion about therapeutic cloning was—and is—much more divided, with some people arguing that the use of human embryonic stem cells for such research was only the first step along a "slippery slope" that would inevitably lead to reproductive cloning, and others insisting that the possible health benefits arising out of stem cell research far outweighed such an unlikely event.

This book is intended as a resource of readers who would like to learn more about the history, technology, and issues related to cloning or who need additional resources to continue their own research on the topic. Chapter 1 provides a broad general introduction to the history of cloning, from its use primarily in the field of agriculture and horticulture to its modern-day applications in animal cloning. Chapter 2 outlines some of the most important issues that have arisen has researchers have extended their studies from plants and simple animals to mammals and humans. The chapter attempts to summarize some of the most common arguments for and against each type of cloning, along with public opinion surveys on the topic, and laws and regulations that have been adopted in the United States and around the world on cloning and its close cousin, stem cell research. The references for each chapter should serve as a valuable resource for additional reading on specific points discussed within the chapters.

Remaining chapters focus on specific aspects of the cloning debate. Chapter 3, Perspectives, provides interested observers and stakeholders with an opportunity to present their views on some specific aspect of cloning. Chapter 4 offers profiles of important individuals and organizations in history and currently who have contributed to the development of and debate over cloning research. Chapter 5 contains some data on cloning and a number of important documents that will be of interest and value to readers who want further references on the topic. Chapter 6 is an annotated bibliography of more than 100 print and electronic resources dealing with cloning. Chapter 7 provides a chronology of important events in the history of cloning. And last is a glossary of important terms in the field.

Cloning

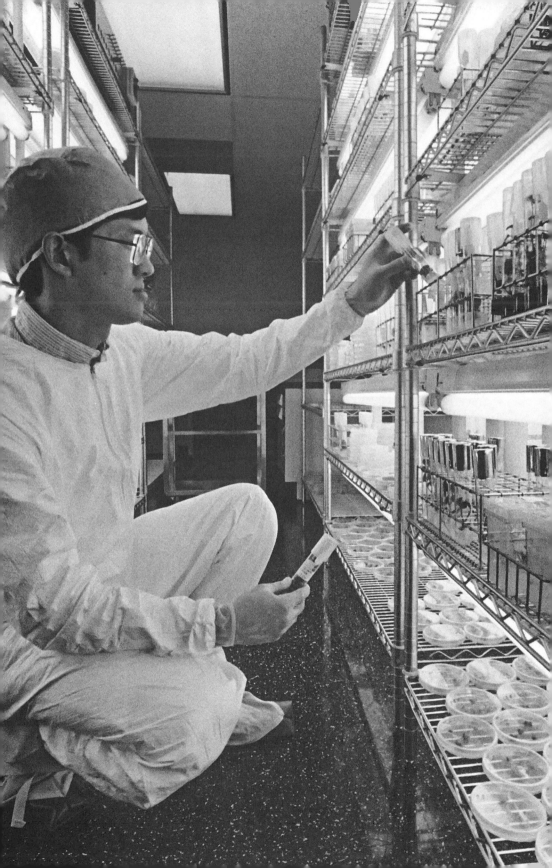

Vivienne had just received bad news. Her doctor had informed her that she had diabetes mellitus. Vivienne was familiar with the disease because her mother, her aunt, two uncles, and her grandfather had all died of the disease. She was not optimistic about her future. Even though it was 1922, doctors knew very little about the cause of diabetes, and they could do almost nothing to treat it. They were able to provide treatment for some of the disease's symptoms, but they had no way of curing or slowing the progress of the disease.

Oscar had just received bad news. His doctor had informed him that he had diabetes mellitus. Oscar knew a little bit about the disease because it seemed to "run in his family." His aunt Vivienne had been diagnosed with diabetes not too many years earlier, and she was very sick then. Researchers in Canada had just discovered the cause of diabetes, the failure of a substance called insulin to metabolize sugar in the body. But, even though it was then 1925, insulin was very difficult to obtain. It had to be extracted from animals that had been slaughtered for meat, and that was a very expensive process. Oscar's family could not afford to buy the insulin he needed to treat his disease. He was not optimistic about his future.

Selina had just received bad news. Her doctor had told her that she had diabetes mellitus. Selina knew a lot about

A scientist in a clean room examines plant tissue in various stages of regeneration, from protoplasts to microcolony to shoot and root formation. (William James Warren/Science Faction/Corbis) *diabetes; it was the beginning of the 21st century, and people knew a lot more about the disease than they did a hundred years earlier. Selina knew that she could make changes in her lifestyle that would reduce her risk for the disease, such as improving her regular diet and exercising more often. Even more important, she knew that she could start on regular treatments with insulin. Her grandmother Vivienne had never had that option because no one knew about insulin in the early 1920s. And her granduncle Oscar did not benefit from insulin treatment because it was much too expensive for his family. But Selina was more fortunate because, in the early 2000s, insulin was being made artificially by genetic engineering in a process known as cloning. The process was relatively simple and inexpensive, and almost anyone who needed insulin could have it at a price he or she could afford. Selina was optimistic about her future.*

Cloning is a process by which a cell, group of cells, or complete organism is produced asexually from a parent cell or organism. (More precise terms for asexual reproduction are *apomixis*, *apogamy*, or *agamogenesis*, terms that refer to methods of reproduction without the fusion of gametes, the male and female reproductive cells involved in sexual reproduction.) The clones produced by such a process are (with a few exceptions) genetically identical to the parent.

The "few exceptions" qualifier used here reflects one of the puzzles with which biologists have been struggling for decades: If a plant or animal reproduces asexually, it simply transfers its genetic material to all daughter cells. In such case, how could the progeny of this process *not* be genetically identical? Yet, the evidence has long been clear that cloned plants and animals are sometimes phenotypically different from their parents (for example, see Smith et al. 2011). For some researchers, the answer to that question appeared to be simple and obvious: environmental factors must be responsible for the development of variant phenotypes among cloned organisms. One authority in the field has written, for example, "The phenotypic variation present within a clone is due to environment only" (Clone and Characteristics of Clones 2014).

Other researchers have been less certain. They have believed that changes must somehow occur in the genome of the parent plant during the process of propagation, producing at least some progeny that are *not* exact copies of the parent. Evidence for that belief has now begun to accumulate. For example, researchers at Oxford University, in the United Kingdom, and King Abdullah University of Science and Technology, in Saudi Arabia, announced in 2011 that they had found that unusually high rates of mutations during asexual reproduction were responsible for the appearance of phenotypes that differed genetically from the parent stock (Jiang et al. 2011; for a more accessible explanation of these results, see Harberg 2011).

Cloning in Nature

Cloning occurs commonly in nature. Perhaps the example of cloning with which most people are familiar involves multiple births that result when a single fertilized egg divides, usually into two new eggs, each of which carries the exact (or nearly exact) genetic material possessed by the original egg. The children born of such a process are *not* genetic duplicates of their parents, although they are copies of the original egg produced by the female and fertilized by the male.

In the most common example of such an event, a fertilized egg divides once to produce a set of identical twins. (Fraternal twins are produced by a different process in which two separate eggs are each fertilized.) Such an event occurs about once in every 250 live births. Much more rare is the production of triplets or quadruplets by this process. In such cases, one or both of the eggs produced by the division of an original fertilized egg divide again, producing a total of three or four eggs identical to the original fertilized egg. When repeated division occurs, either identical triplets or identical quadruplets are born, all of whom have genomes identical to the original egg. The only known instance of identical quintuplets who survived childhood were the Dionne quintuplets, born on May 23, 1934, in the town of Corbeil in northern Ontario (Wright 1994/1995).

Another form of cloning that occurs in nature is parthenogenesis, a form of reproduction that occurs when an unfertilized egg spontaneously divides, producing two or more exact copies—clones—of the original egg. Parthenogenesis occurs naturally in many plants and a wide variety of animals, including invertebrates such as nematodes, water fleas, scorpions, aphids, bees, some parasitic wasps, and a number of so-called stick insects and walking sticks, as well as a small number of vertebrates, including certain fish, amphibians, reptiles, and birds. In all such cases, the offspring produced by parthenogenesis are clones because they carry a genetic composition identical to that of their parent.

Parthenogenesis has long been a topic of considerable interest to biologists who are interested in the ways in which this type of asexual reproduction differs from the more familiar use of sexual reproduction, common among both plants and animals. For example, research suggests that many species that use parthenogenesis for reproduction do not do so in all cases but alternate between asexual and sexual reproduction in succeeding generations. The evolutionary reason for such a pattern may well be the need for the introduction of genetic diversity that contributes to a species' long-term survival and that is not generally available from asexual reproduction in and of itself (Vrijenhoek 1998).

From a strictly numerical standpoint, the most common form of reproduction on Earth is probably asexual reproduction, the production of offspring from a single organism. For example, a common form of reproduction among bacteria is fission, in which a single bacterial cell divides into two new cells, both of which have an identical genetic composition to the parent cell. A parent cell donates its genetic material in the form of DNA (deoxyribose nucleic acid) to each of the new cells produced from it. Unless there is an error in reproduction (such as might occur as the result of exposure to radiation or certain chemicals), all daughter cells produced during asexual reproduction are genetically identical to the parent cell; they are all clones of that cell. Most members of the domains of Eubacteria and Archeabacteria as well as the eukaryotic Protista reproduce asexually. Many plants and some higher animals also reproduce asexually, as by parthenogenesis or other asexual modalities.

Some other types of asexual reproduction include budding, vegetative propagation, spore formation, and fragmentation. Budding occurs when a parent organism develops an outgrowth (bud) on its outer surface. The outgrowth continues to develop until it becomes a mature organism, at which point it detaches itself from that organism and begins living an independent life. Since the new organism is produced by the removal of parts of the parent organism, it is genetically identical to that organism and is a clone of that organism. Budding is the most common form of reproduction in a number of simpler animals, such as the hydra, certain corals and sponges, and a number of flatworms. (For a video demonstration of budding in hydra, see Budding in Hydra, 2010.)

An interesting variation of budding is called endodyogeny, in which two daughter cells are produced within a parent cell. When the daughter cells reach maturity, they devour the mother cell and begin independent life cycles. Endodyogeny occurs among a number of parasites, such as *Toxoplasma gondii* (Hu et al. 2002).

As with other forms of reproduction, budding can take a number of other forms. For example, the somewhat rare event known as endopolygeny involves the budding of multiple daughter cells within a parent cell, which then, as with endodyogeny, is consumed by its offspring. This modality is also used by parasitic organisms, such as *T. gondii*. (For more details about this process, see Ferguson 2009.)

Some forms of asexual reproduction, such as fragmentation and spore formation, occur especially commonly, but not exclusively, in plants. When such forms of reproduction do occur in plants, they may be collectively known as vegetative reproduction, vegetative propagation, or vegetative cloning. In the process of fragmentation, a portion of the parent organism is detached, either accidentally or naturally, and that fragment then begins a new life cycle independent of the parent organism. Many different types of fragmentation are possible. For example, some plants produce adventitious structures, which then become detached and grow into independent plant. The term *adventitious* refers to structures that appear in unusual places. For example, the common house plant, *Kalanchoe daigremontiana*, gets its common name of "mother of thousands" because of its mode of reproduction. The plant develops tiny plants (called *plantlets*) on its leaves, which, over time, drop off and become rooted in starting their new life cycle. Of course, each plantlet and each new daughter plant is an exact genetic copy, and hence a clone, of the mother plant (The Weird and Wonderful Ways Plants Reproduce 2014). In many cases, these adventitious structures develop when a plant is injured or otherwise loses a part of its structure. The development of adventitious structures then provides a mechanism by which the plant can ensure the continuation of its DNA to future generations.

Two other forms used by plants for propagation by fragmentation are structures known as rhizomes and stolons. A rhizome is a modified underground stem that can develop into the root system of a new plant. The rhizome may remain connected to the parent plant, eventually becoming part of a colony of plants interconnected through a complex root system. Or the rhizome may become detached from the mother plant, forming the basis for a new plant, genetically identical to its parent (Sumac Control 2014). Some examples of plants that reproduce by means of rhizomes include asparagus, bamboo, Bermuda grass, Cannas, Chinese lantern, ginger, hops, irises, Lily of the Valley, sympodial orchids, Venus flytrap, and Western poison oak.

Closely related to the rhizome is a stolon, also a stem that grows out of a parent plant, either on the surface of the ground or just below it. Probably the most familiar example of a stolon-producing plant is the strawberry. New plants develop at the apex of the stolon, which then mature and continue to grow even if the stolon is lost or destroyed (Kraehmer and Baur 2013). Other plants that reproduce by means of stolons include the water hyacinth, *Acaena, Saxifraga stolonifera*, the white potato, and the common garden and house plant, hen and chickens.

Fragmentation is also fairly common among a number of animal species. In this process, an animal reproduces by losing a portion of its body, which then regenerates to form a new organism. Fragmentation sometimes occurs as a normal and natural part of an animal's life cycle; it is programmed to detach a portion of its body, which then regenerates to produce a complete new organism in a process known as *autotomy* (from the Greek for "self-severing"). In other cases, the process occurs as the result of injury to some portion of an animal's body, during which the severed portion regenerates into a complete new organism and the parent organism repairs the damage to its own body.

Fragmentation is a common form of asexual reproduction among sponges, corals, annelids, and echinoderms. For example, the process is the primary means of reproduction among some species of starfish (class Asteroidea) that can fragment and regenerate from fragments no larger than about a centimeter in length. This fact has long had an interesting commercial implication since oyster fishermen have traditionally cut starfish into segments and thrown them back into the ocean to stop them from preying on oyster beds. This action is counterproductive, however, since each of the starfish segments eventually regenerates, forming a new organism and increasing the number of oyster predators many times. Recent research studies have also shown that fragmentation is the primary and critical method by which some species of corals reproduce and contribute to the animal's survival in some parts of the world (Fautin 2002; Highsmith 1982; Starfish (Seastars) Regenerating Their Arms 2014).

Yet another form of asexual reproduction in plants involves the formation of spores, a unicellular particle capable of developing into a new individual. Most spores are capable of surviving severe environmental conditions that would otherwise be fatal for an organism, such as unusually hot, cold, or arid conditions. Most spores are produced by meiosis and are an element of the sexual development and reproduction of a plant. But some plants produce spores by simple mitosis, or cell division, resulting in spores whose genetic makeup is identical to the parent organism. These spores are known as *mitospores*, to recognize the method by which they are formed. They leave the parent plant and may remain in the environment for months or years before beginning to grow and develop, forming new plants that are clones of the parent plants. Mitotic sporogenesis occurs most commonly among certain species of fungi and algae, perhaps most prominently by the conidial fungi *Aspergillus* and *Penicillium* (Differences between Mitospores and Meiospores 2014).

Cloning in Horticulture: Field Methods

Cloning has been an essential part of horticulture for many centuries. Historical records suggest that humans had learned how to produce clones by vegetative propagation at least as long ago as the first century CE. The Roman author and natural philosopher Pliny the Elder (23–79 CE), for example, described a number of plants, including apples and quince, that were being propagated for very specific characteristics during his lifetime (Janick 2005). A number of cultivars familiar to contemporary consumers reflect the staying power of plants that have been maintained through vegetative propagation because of their unique or very specialized qualities, such as the clementine mandarin, "Kerman" pistachio, and Dwarf Cavendish banana. The usual approach was to use hybridization to produce a plant with some particular desirable quality or qualities and then to continue that cultivar's lineage through asexual reproduction. Two well-known examples of this process are the Bartlett (Williams) pear and the Delicious apple. The former is thought to have originated in Great Britain in about 1770 in the orchard of a man by the name Williams, accounting for one of its modern names. It was then transferred to the United States where it was grown on an estate in Dorchester, Massachusetts, owned by Enoch Bartlett, accounting for its other current name (The History of Bartletts 2014). A similar story can be told about an apple that was once the best-selling apple in the United States, the Red Delicious. Developed on an Iowa orchard in about 1880, it grew to become the most popular eating apple in the United States a century later, but then lost its appeal to the American public in a matter of years (Higgins 2005; Seabrook 2011).

These examples are not meant to suggest that once horticulturists produce a successful clone that the plant never again changes its genetic character. As with all other aspects of agriculture, men and women in the field are constantly attempting to improve their product: improve taste, make production less expensive, find a way to make their product last longer, or bring about some other changes in a long-standing favorite cultivar. The Bartlett pear or Red Delicious apple available in stores today is by no means the same product that was originally developed a century or two ago. Horticulturists have refined and improved their versions of these products, often by hybridization, and, when a superior product has been found, it becomes the new clone that goes to the marketplace. (For more detail on this process, see Kumar 2006, 32–37.)

A number of methods are available for the production of plant clones by means of vegetative propagation. These methods are sometimes divided into two general categories, those that are used in the field and those that are used in the laboratory. Under the former category are cuttings, grafting, budding, and layering. The most common type of vegetative propagation used in the laboratory is tissue culture.

Cutting is the most common type of vegetative propagation. It involves the removal of a portion of the parent plant, which is then placed into some type of medium that encourages the generation of a new plant that is (generally) genetically identical to the parent plant. (Circumstances may result in the formation of a clone that eventually reverts to an earlier version of the parent plant and is not, therefore, genetically identical to that parent plant.) Cuttings usually involve the removal of some portion of a stem, leaf, or root, which is then planted in an appropriate medium. Specific procedures are used for each type of cutting that is taken and planted. For example, a stem cutting can be taken at the end of the stem (a tip cutting), along the length of the stem itself (a medial cutting), or at the base of a leaf or bud (bud cutting). A leaf cutting may involve the removal of an entire leaf, with or without its petiole, or a portion of the leaf blade. Wherever the cuttings are taken, they are then planted in a suitable medium, usually determined by the type of plant being cloned. Some examples of media used are sand, peat, perlite, vermiculite, pea gravel, loamy soil, or some combination of these materials. (For more information on vegetative propagation, see any textbook on horticulture or Bryant 2006; Plant Propagation: Asexual Propagation 1998; Reif and Ball 2009; Resource Book on Horticulture Nursery Management 2014.) Some examples of plants that can be propagated by cutting technology (as well as other technologies) are given in Table 1.1.

Grafting is a procedure by which parts of two separate plants are joined to each other so that they can grow as a single new plant. The portion that is removed from a parent plant is known as the *scion*, while the plant to which the graft is attached is called the *rootstock*, or simply *stock*. In a grafted plant, only the portion that grows out of the scion is a clone of the parent plant; the rootstock retains its original genetic characteristics. The essential requirement of grafting is that the scion be able to grow unto the rootstock because some combinations of plants will not develop a common bond. A variety of grafting techniques are available, including bark, bridge, cleft, in-arch, saddle, splice, and whip and tongue. The various types of grafts get their names from the way the scion is inserted into the rootstock. A cleft graft, for example, is made by cutting vertically into the stem of the rootstock and inserting the scion into the cleft. By contrast, the bark graft involves the insertion

Table 1.1   Examples of Plants Grown by Vegetative Propagation

Method		Plants
Cuttings		
	Stem	*Acer* (maple) *Ananas* (pineapple) Cannabaceae (hemp, hops) Manioc (cassava) *Olea* (olive) *Rubus occidentalis* (blackberries) *Vanilla* (vanilla) *Vitis* (grapes)
	Leaf	*Episcia* (flame violets) *Hoya* (waxplant) *Peperomia* (radiator plant) *Saintpaulia* (African violets) *Sedum*
	Root	*Campsis* (trumpet vine) *Ficus* (fig) *Phlox* (phlox) Rosaceae (roses) *Syringa* (lilac) Sumac (sumac)
Grafting		Many varieties of trees, such as apple, ash, beech, cedar, cherry, citrus varieties, horse chestnut, maple, and spruce
Budding		Woody fruit trees in general
Layering		Azalea *Clethra* *Forsythia* *Hydrangea* Quince *Rhododendron* *Spirea* Weeping willow *Weigela*

of the scion into a cut made between the bark and the wood (core) of the plant (Grafting and Budding Nursery Crop Plants 2014; Kumar 2014; Reddy and Rao 2010, Ch. 3; Vegetative Propagation Techniques 2007).

Budding is a specialized form of grafting in which the scion is a single bud. Therefore, the techniques used in the process are generally similar to those used for grafting except a much smaller piece of plant material is used in budding than in grafting. The new plant formed from a budding procedure is a genetic clone of the parent plant from which it was taken (Grafting and Budding Nursery Crop Plants 2014; Reed 2007; T or Shield Budding 2014).

Layering is a plant propagation technique that is thought to date back at least to the Roman Empire, and probably much earlier (Mudge et al. 2009). As with other vegetative propagational techniques, layering is used with plant species that cannot easily be reproduced by other means. Two forms of layering are most common. In air layering, a cut is made on the stem of a plant that removes the bark and cambium. The plant is "wounded" by this operation, stimulating it to produce new plant material as a way of repairing the damage caused by the plant. The wounded area is then covered with a material that can be kept damp, such as a mat of sphagnum moss enclosed in a plastic cover. Over time, the stem generates new roots at the injured region, roots from which a new plant—a clone—begins to grow. After the new plant appears to be mature enough to survive on its own, the section of the stem below the plant is removed from the parent plant and placed into the ground, where it begins to grow on its own.

Tip layering is somewhat similar to air layering except that it involves bending the end of a branch from a parent plant over until it touches the ground. The tip is then wounded, as with air layering, covered with soil, and kept damp. Eventually the parent plant begins to heal the wounded area by the production of new roots. The new plant formed from those roots then begins to grow in place as a clone of the parent plant (Janne 2014; Reed 2007).

It should be noted that the vegetative propagation techniques described here have often been used for purposes other than the production of plant clones. Botanical historians know, for example, that these techniques have also been used for at least two millennia for the production of artistic plant displays, such as the complex arbosculptures still seen in many arboreta, public parks, and private homes (Mudge et al. 2009).

Plant Tissue Cell Cultures

Cloning of both plants and animals in laboratory settings is also a common procedure in today's world. Unlike field methods of asexual reproduction, which have been practiced for centuries, the most common type of laboratory cloning, tissue culturing, has been practiced successfully for little more than a century. Tissue culturing (also known as *cell culturing*) is the practice of removing small pieces of a plant or animal and then transplanting those pieces into environments that promote their regeneration to complete clones of the parent plant or animal. This process is sometimes known as *explantation* ("to *explant*" a material), and the material taken from the parent plant or animal is also known as an *explant*.

The notion that a plant or animal could be regenerated from a small piece of tissue or even a single cell dates to the 1830s, when the cell theory was first proposed by German biologists Matthias Schleiden and Theodor Schwann. An important implication of that theory was that plants and animals should be able to reproduce themselves from segments taken from their own bodies because the cells of which they are made are autonomous, that is, capable of reproducing on their own to make new organisms genetically identical to themselves. (A good general introduction to and early history of studies on regeneration is Birnbaum and Alvarado 2008.)

Although this idea seemed obvious to those biologists who understood and adopted the cell theory, actual experiments in the regeneration of a plant or animal from tissue or individual cells progressed only very slowly. The first such experiments were those conducted by Austrian botanist Gottlieb Haberlandt in 1902. Haberlandt attempted to generate a number of different plants by taking small cuttings from plant bodies and transplanting them into nurturing media consisting of Knop's salt solution. Knop's is a traditional nutrient-rich solution used in growing plants, consisting of potassium nitrate, magnesium sulfate, dibasic potassium phosphate, and calcium nitrate. Haberlandt used a variety of tissues, such as palisade cells from the leaves of *Lamium purpureum*, glandular hair of *Pulmonaria*, and pith cells from petioles of *Eicchornia crassiples* (Rai 2007; Haberlandt's description of his experiments constitutes the first section of Laimer and Rücker 2003, 1–24).

Haberlandt's experiments were a "success" in the sense that the transplanted tissue survived for a number of months and accumulated starch, indicating that they were, indeed, "alive." However, they ultimately failed to replicate so that the experiment overall had to be considered a failure. Probably more to the point, however, was the description that Haberlandt provided of his work and the impetus that the work had on other researchers. In fact, in spite of the failure of his efforts to regenerate plants by tissue cloning, he is generally regarded today as "the father of plant tissue culture" (Krikorian and Berquam 1969; Thorpe 2007, 10).

Perhaps Haberlandt's greatest contribution to the progress of tissue culture was his belief in a property he called *totipotency*, a property he thought was possessed by all cells in a plant. The term totipotency refers to the ability of an individual cell to develop into any type of specialized cell. Human totipotent embryonic cells are capable, for example, of developing into a variety of specialized cells, such as muscle, nerve, bone, skin, blood, or other cells. Based on this belief, Haberlandt saw no reason that complete plants could not be produced by tissue culturing beginning with individual cells or groups of cells from any part of a plant. Presciently, he once wrote that "I believe, in conclusion, that I am not making too bold a prediction if I point to the possibility that, in this way, one could successfully cultivate artificial embryos from vegetative cells" (Sussex 2008, 1191). Haberlandt's observation could hardly be a better description of the status of plant cell culture today.

Haberlandt's hypothesis proved to be a powerful impetus for research on tissue culture experiments in coming decades. A number of researchers attempted to clone mature plants beginning with only small amounts or a few cells of a parent plant. Although Haberlandt believed that *all* cells in a plant are totipotent, subsequent researchers tended to choose cells and tissues that seemed most likely to be able to propagate on their own. For example, in 1908, German botanist Siegfried Simon attempted to regenerate poplar plants from stems of parent plants. He was able to produce buds, roots, and calluses in his experiments, although none of these structures then went on to become complete plants (Gautheret 1983; Vasil 1985, 8–9).

For this line of work, Simon is sometimes mentioned as the "father" of callus research that became so important in the later development of plant tissue culturing. A callus is a mass of cells that develops on a plant at the site of and following an injury to the plant. Botanists had been intrigued by the character of plant calluses for years and eventually discovered that they consisted of undifferentiated cells that were totipotent, capable of developing into any type of cell needed to repair the damaged region of the plant. As a consequence, calluses became an important material in later tissue cell culture experiments (Basics of Plant Tissue Culture 2014).

Another approach growing out of Haberlandt's research challenge was the use of embryos for the development of new plants. Embryos are an obvious place to begin in the process of cloning plants because, of course, all plants begin their life cycle with embryos in one form or another. In 1904, for example, German botanist Emil Hannig removed nearly mature embryos from seeds of various species of crucifers (*Raphanus caudatus*, *R. landra*, *R. sativus*, and *Cochlearia danica*) and grew them to maturity in a solution containing mineral salts and sugars. The plants all grew to maturity but, again, did not reproduce (Raghavam 2003). Nonetheless, Hannig's work is generally regarded as the origin of embryonic tissue culture (Jourdin 2010).

Yet another approach to the use of plant materials in tissue culturing involved starting with meristematic cells. The term *meristem* refers to undifferentiated tissue found at parts of a plant that are undergoing growth, such as roots and stem tips. In 1922, for example, two researchers, Walter Kotte, a student of Haberlandt, and American botanist William J. Robbins, reported on their research in this area. The researchers used root tips from corn, cotton, and maize plants to produce young plants that survived for short periods of time, but ultimately stopped growing and never became independent new plants (Kavanagh and Hervey 1991; Robbins 1922; Sussex 2008; Kotte's 1922 paper is reproduced in Walter Kotte: Wurzelmeristem in Gewebekultur 1922).

In spite of efforts such as these, progress in plant tissue culture took place only very slowly in the three decades following Haberlandt's outline for research in the field. Most experiments had to be regarded as failures because they failed to meet at least one of the three essential criteria for successful tissue culture experiments. These criteria were not even specifically described until the 1930s but involved the creation of a mass of cells (1) that was capable of unlimited growth, that is, could continue to reproduce on its own for an unlimited period of time; (2) that was capable of remaining undifferentiated during that period of time, that is, continue to consist of totipotent cells; and (3) that would eventually become differentiated so that it could reproduce all the elements of a complete mature plant. (The clearest statement of at least part of this mission was offered by American botanist Philip R. White. For example, see White 1936, 1939.)

Progress in plant tissue culture experiments was slow in the decades following the challenges posed by Haberlandt in 1902. Historians have pointed to two main reasons for this fact. First, experimenters sometimes chose plant materials to work with that were inherently difficult to grow and propagate. In a general review of developments in the field of plant tissue culture in 1951, for example, White noted that many experiments had failed simply because researchers had chosen inappropriate plant materials with which to work (T. Murashige and F. Skoog 1962; White 1951). As researchers solved this problem, the second issue became more obvious: the need for a suitable growing medium.

Early researchers often chose fairly simple platforms on which to grow plant tissues, usually a mixture of mineral salts and sugars. Such a choice made sense because relatively little was known about the nutritional needs of growing plants *except* for the fact that they required inorganic ions, such as those of calcium, nitrogen, phosphorus, sodium, and potassium, along with glucose and fructose. Over time, however, they began to try a variety of additives to this basic mixture in an effort to improve the nutritional setting for plant materials. These additional materials included coconut milk, fruit juices, malt extract and yeast extract (YE), and hydrolyzed casein (George, Hall, and de Klerk 2007, 115 *et seq.*). Most of these medium additives improved the growth of tissue cultures because, unbeknownst at the time to researchers, they each contained a substance that either was essential for or contributed to the growth and reproduction of plants.

During the 1930s, a number of efforts were made to identify the specific components of the natural products being used as medium additives. An important breakthrough occurred in 1934 when White found, first of all, that tomato root tips not only grew well but also could be subcultured in a new medium that contained YE. He then went on to refine the experiment by substituting a component of YE, vitamin B, demonstrating that the vitamin was essential to plant growth and reproduction (White 1934; the specific medium used by White consisted of a mixture of inorganic salts along with a 2 percent solution of sucrose and 0.01 percent solution of YE; Sussex 2008). This result was especially significant because virtually nothing was known at the time about the role of vitamins (about which little in general was then known) in promoting plant growth. From that time on, researchers understood that plant tissue culture mediums required the inclusion of one or more B vitamins, such as pyridoxine and thiamine, to achieve normal growth.

An equally important discovery occurred in 1926 when Dutch botanist Fritz Went discovered the first auxin, indole-acetic acid (IAA). Auxins are hormones that promote and regulate the growth of plants. The discovery of the role that auxins play in plant growth provided at least a partial explanation as to why researchers had been so unsuccessful in maintaining plant tissue cultures with simple platforms consisting of mineral salts and sugars; those platforms also required auxins to encourage the continued survival, growth, and reproduction of the plant materials being studied.

One of the first demonstrations of that fact came in 1934 when French botanist Roger Jean Gautheret added IAA (along with vitamin B) to the medium in which he was culturing cambium cells taken from a variety of tree species. Gautheret found that the cells growing on this medium promoted the proliferation of the cambial cells, resulting in a synthetic callus that appeared to have an essentially infinite lifetime, one of the requirements of a successful plant tissue culture. Some historians have suggested that Gautheret's experiment resulted in the production of "the first true plant tissue cultures" (Thorpe 2013, 2).

Over the next two decades, researchers continued to add to their armamentarium of techniques for use in plant tissue culture experiments. For example, three research groups in 1939 announced almost simultaneously that they had developed cultures that appeared to be essentially immortal. That is, the cultures met the two essential features of a tissue culture, namely that they continued to survive and reproduce as a genetically identical population. In one case, two French workers, Gautheret in Paris and Pierre Nobécourt in Grenoble, reported success in developing immortal cell cultures from tissue taken from carrot tap roots grown on mediums containing IAA (Gautheret 1939). At about the same time, White reported

similar success in working with cells taken from a hybrid tobacco plant *(Nicotiana glauca* × *N. langsdorffii)* treated with IAA
that also appeared capable of reproducing perpetually without
variation (White 1939). (Gautheret and Nobécourt appeared
to have been unaware of White's somewhat earlier work.)

The approach that many experimentalists used was first to
extract cells from a parent plant and place them into a medium
with known nutritional resources needed for the growth of the
cells. Next, the cells might need to be dedifferentiated. That
is, if they were cells taken from a root or leaf, they had to be
restored to a more primitive, totipotent, stage that would allow
them to develop into a callus or callus-like culture. Evidence
seemed to suggest that dedifferentiation depends upon and is
a function of the presence of auxins and, perhaps, other plant
hormones in the medium (Birnbaum and Alvarado 2008, 5–6;
Ikeuchi, Sugimoto, and Iwase 2013). After the cell culture thus
formed had been proved to be totipotent and immortal, it then
had to be transformed once more to a differentiated state in
which it could develop into a complete plant with all of the
essential elements that such an organism requires. Researchers
thus had to find ways of reversing the process of dedifferentiation and force the callus to begin developing roots, stems, and
other plant parts.

A key development in dealing with this challenge was the
work of Swedish-born American plant physiologist Folke Skoog
and his colleagues at the University of Wisconsin after World
War II. They found that that *quantity* of plant hormones available in a tissue culture was a potent factor in determining
whether or not differentiation (or dedifferentiation) occurs and,
if it does, what effects those hormones may have. In a classic
1957 paper, they explained how a proper balance of auxin and a
second hormone, kinetin, controlled whether or not differentiation even occurred and, if it did, whether shoots or roots formed
in the callus (Amasino 2005; Skoog and Miller 1957).

Just as the 1957 Skoog paper was being published, two
other important studies were reported that represented the

culmination of a half century's work on plant tissue culture, experiments that some historians regard as the starting point of modern cloning research. Both of these studies involved the production of new embryo from somatic cells, a process known as *somatic embryogenesis*. In this process, fully developed and differentiated cells dedifferentiate to form embryonic cells from which fully developed and differentiated plants are then able to grow. (This process differs from *zygotic embryogenesis*, in which male and female germ cells combine to form the embryo.) In both of these experiments, tissue from mature carrot (*Daucus carota*) plants were macerated and then either maintained in a suspension culture (Steward, Mapes, and Smith 1958) or converted into a callus (Reinert 1959). (A suspension culture is a system in which individual cells or groups of cells are held in suspension in a nutrient mixture.) Under the proper conditions, these individual cells and groups of cells dedifferentiated to become embryonic cells, which then began to grow into fully developed carrot plants. These experiments were significant because they demonstrated the fact that plants could be cloned not only from roots, stems, buds, and other plant parts, as had been done for centuries, but also from individual plant cells, which apparently contained all the genetic information needed for the production of complete new plants. A number of historians of plant tissue cell culture have pointed out that other researchers were in the process of obtaining results similar to those of Steward and Reinert at about the same time, suggesting that credit for being named as "Father of Cloning" could easily be distributed among a number of individuals. (See, for example, Dodds and Roberts 1985, Table 11.1, 123; Krikorian and Simola 1999; Muir, Hildebrandt, and Riker 1958.)

Animal Cloning

Interest in the cloning of animals arose at about the same time as that for the cloning of plants, in about the 1880s. An important impetus for this line of research was a theory of heredity proposed independently and at almost the same time by German biologists August Weismann and Wilhelm Roux (although the theory tends to be associated more commonly with Weismann than with Roux). According to this theory—which has come to be known as the germ plasm theory—sexual reproduction occurs when an egg and a sperm combine to form a zygote, which then contains genetic information from both parent cells. As this cell begins to divide, that genetic information is divided upon between the daughter cells: half goes to one cell in the first stage of division, and the other half goes to the other cell. This theory means that the amount of genetic information is reduced during each stage of cell division, with some cells retaining all the genetic information needed to make, for example, a liver cell; some retaining the information needed to make a skin cell; and so on.

One implication of this theory is that genetic information is transferred in only one direction in the process of growth and differentiation: from germ cells to somatic cells (or to other germ cells), but never from somatic cells to germ cells. That is, the germ plasm theory insists that it would be impossible use a somatic cell (such as a skin or muscle cell) to produce a new animal because that cell lacked most of the genetic information needed for that process. (For a complete description of Weismann's theory, see Weismann 1893. Roux's theory is described in Kearl 2013b; his original paper on the theory is found in Roux 1883.)

Early research appeared to support the Weismann–Roux theory of development. Roux himself carried out some of the most important of those studies. He announced the results of perhaps the most influential of this research in an 1887 paper, "Beiträge zur Entwickelungsmechanik des Embryo" (Roux 1887). In that research, Roux used a hot needle to destroy one of the cells in a two-cell embryo of the green frog (*Rana esculenta*). He found that the procedure resulted in the formation of a half embryo that would develop into a complete embryo consisting of only half of the parts of a complete embryo, a result he actually observed. He concluded from these results that the Weismann–Roux theory was correct and that genetic information was lost at each stage of the development of an embryo (Gilbert 2014, online version; McKinnell 1985, 24–26).

As it turns out, Roux was incorrect in the interpretation of his results, as a number of other experiments were soon to demonstrate. One series of such experiments has become one of the most famous bits of research in the history of cloning. In the early 1890s, German biologist Hans Driesch began a series of experiments that closely matched those conducted by Roux. The major difference was that Driesch chose to work with two-cell embryos of sea urchins (*Echinus microtuberculatus*) rather than frogs. Because the sea urchin embryos were much smaller than those of frogs, Driesch had to use a different method for separating the cells in an embryo; the use of a hot needle would have destroyed the whole embryo, not just one of its cells.

Driesch chose to simply shake a suspension of urchin embryos until the two cells of which they were made broke apart. He then observed how the individual cells thus formed would develop (or not). He was amazed to discover a result very different from that obtained by Roux, namely that individual embryonic cells were capable of developing into complete sea urchin organisms, although they tended to be smaller than animals that developed normally from two-cell embryos. He repeated a number of variations on this experiment, using four-, eight-, and more developed embryos, varying the process of separating cells in an embryo and working with a number of other invertebrate species, including echinoderms, ctenophores, and ascidians (Gilbert 2014, online version; Kearl 2013a; Sunderland 2013).

At the dawn of the new century, an experimentalist was to appear on the scene who was to provide some of the best new information about the process of animal cloning, Hans Spemann. Spemann was a German embryologist who was eventually to win the 1935 Nobel Prize for Physiology of Medicine for his research on the mechanisms by which a fertilized cell develops into the specialized cells of which a mature organism is made. He reported on the earliest of these studies in his 1902 paper "Entwickelungsphysiologische Studien am Triton-Ei" ("Physiological Development of the Triton Egg"). In these early experiments, Spemann chose to work with salamander eggs because they are large, transparent, easy to manipulate, and easy to incubate. (Triton is an early name for aquatic salamanders.)

As suitable as salamander eggs appeared to be for his studies, Spemann faced one difficult challenge. He was unable to separate the two cells of an early embryo simply by shaking, as Driesch had done, because the embryo was covered with a sticky cover that protected it from just such damage. Spemann was able to invent an ingenious solution to this problem, however. He used a strand of hair from his newly born daughter to tie off two parts of the embryo from each other, allowing him to observe the development of each of the halves produced in this way. He was overjoyed, but a bit surprised, to find that each of the two cells grew into a normal adult salamander, in agreement with Driesch's research, but in opposition to Weismann and Roux's predictions. (Spemann called his procedure *twinning*.)

Spemann then went one step further. He repeated his earlier experiment, this time, however, using a fertilized egg *before* it had undergone its first division. He was interested in finding out what part of that fertilized cell was responsible for the later development of the embryo. In this second experiment, then, he tied off the fertilized cell in such a way that the nucleus was in one half of the egg and the rest of the cell, in the other half of the egg. He noted in this experiment that the portion of the egg containing the nucleus continued to develop normally into an adult salamander, while the half lacking the nucleus underwent essentially no change.

More than two decades later, Spemann moved on to an even more advanced and sophisticated stage of his research. He began, as before, by tying off a fertilized egg at the two-cell stage and allowed the zygote to develop over a period of hours. He found, as before, that the half of the embryo containing the nucleus—which had reached either the 8- or 16-cell stage—was, at that point, in the process of developing normally into an adult salamander. The half containing no nucleus had not differentiated. Spemann then loosened the hair constricting the embryo, allowing the nucleus from one of the developing differentiating cells to slide over into the opposite half of the embryo. After a few more hours, Spemann observed that the previously undifferentiated half of the embryo had begun to grow and differentiate. It had begun developing in the same way as the other half, except that its development was delayed in time by an amount during which it contained no nucleus. (Anderson 2004; Animal Development: From Genes to Organisms: Experiment Links 2014; Beetschen and Fischer 2004, 609. For illustrations of these experiments, see The History of Cloning 2014.) At this point, the evidence had become fairly clear that the germ plasm theory of Weismann and Roux was incorrect and that clones of both plants and animals could be produced from a single embryonic cell.

The success of these experiments caused Spemann to look even further into the future. By 1930, he had demonstrated that embryonic cells after three or four divisions were still totipotent, capable of developing into normal adult organisms. But, he wondered, for how long would that situation continue: after 5 divisions, 10 divisions, or even more? That is, would it be possible to isolate a fully developed adult cell of an organism and somehow clone that cell as he had done with 8- and 16-cell embryos?

Spemann described in his 1938 book, *Experimentelle Beiträge zu Einer Theorie der Entwicklung (Embryonic Development and Induction)*, a line of research that he referred to as a "fantastical experiment." In such an experiment, Spemann suggested, one would remove the nucleus from an unfertilized egg and replace it with the nucleus from a differentiated embryo. It was his view, Spemann said, that

Decisive information about this question may perhaps be afforded by an experiment which appears, at first sight, to be somewhat fantastical. . . . Probably the same effect could be attained if one could isolate the nuclei of the morula and introduce one of them into an egg or an egg fragment without an egg nucleus. . . . This experiment might possibly show that even nuclei of differentiated cells can initiate normal development in the egg protoplasm. (Spemann 1938, 211; as quoted in Lensch and Mummery 2013, 9)

Unfortunately, Spemann knew of no way to carry out such an experiment, and it was more than a decade before any other researchers worked out the protocol for this type of research and carried out what is now the basic model for cellular cloning in modern science.

Although Spemann had provided the theoretical overview of a cloning experiment, he was unable to proceed any further because he (and everyone else in the field) lacked the techniques required to perform such an experiment. Just 15 years later, however, sufficient progress had been made to allow the "fantastical experiment" to become a reality.

To reach this point, three basic technical problems had to be solved. First, a method had to be devised to remove the nucleus of an egg (to *enucleate* the egg) that was to become the host cell in the procedure, without actually destroying the egg. Second, a method had to be found for removing the nucleus of a second cell, the donor cell, which was then to be inserted into the host egg. Third, a method had to be developed for transferring the donated nucleus to the host egg, again without damaging either the transplanted nucleus or the host egg. By 1952, American researchers Robert Briggs and Thomas J. King had essentially solved all three of these problems. They were ready to attempt the "fantastical experiment."

For this experiment, they used the Northern leopard frog (formerly *Rana pipiens*; now *Lithobates pipiens*). They began by activating and removing the nucleus from blastula cells of the frog by simply pricking the egg with a clean glass needle. This act caused the nucleus to move into a position from which it could be removed from the egg by drawing it out with a micropipette. They then removed the nucleus from a second blastula cell by a similar process of drawing it out with a micropipette. Finally, they inserted the new nucleus into the host cell and allowed the cell to grow and differentiate. The result of the procedure was the development of a complete and healthy tadpole (Gilbert 2014; see especially Figure 4.6; the original Briggs and King paper is at Briggs and King 1952; for an animation of the procedure, see Melton 2006 or What Is Cloning 2013).

Briggs and King had clearly solved one problem of embryogenesis; they had showed that a cell taken from an early-stage embryo was totipotent, capable of developing into a complete organism consisting of differentiated cells. But was that also true of more developed cells, cells present in 16-cell embryos and their more complex descendants? After a number of studies using more advanced structures such as these, Briggs and King came to a conclusion about this question. They decided the answer was "no." Attempts to clone complete organisms from intestinal or other types of specialized cells failed. It appeared that once a cell had become differentiated, something happened to its genetic makeup that prevented the development of an adult organism (Gurdon and Byrne 2003).

One person who had reservations about this conclusion was British zoologist John B. Gurdon. Gurdon suspected that the reverse of Briggs and King's conclusion might be true and that adult differentiated cells might, indeed, dedifferentiate and become totipotent, meaning that they could then be reprogrammed to grow into new adult organisms. Some of his earliest experiments to confirm this hunch were unsuccessful, partially successful, or of dubious success. However, he was eventually able to demonstrate that, indeed, intestinal and epithelial cells taken from adult African clawed frogs (*Xenopus laevis*) could be used in transplantation experiments to produce normal, adult, fully developed frogs. His results, subject to severe question and scrutiny for a number of years, were eventually accepted by the scientific community, and the question of cloning adult animals from differentiated cells was taken as solved (Gurdon 2009; a list of Gurdon's papers on this series of experiments can be found at Maayan 2013, Sources).

Onward and Upward

In 2012, John Gurdon was awarded a share of the Nobel Prize in Physiology or Medicine for his 1962 "proof of concept" and subsequent experiments on the cloning of animals from mature, differentiated cells. A "proof of concept" experiment is one in which some fundamental theory about a subject is shown to be true. An example is the experimental flight conducted by the Wright brothers at Kitty Hawk, North Carolina, on December 17, 1903. That flight demonstrated that heavier-than-air flight was possible. In Gurdon's case, his experiments showed that Spemann's "fantastical experiment" contained a critical kernel of truth, that mature cells could be used for the cloning of adult animals similar to those from which the cells originally came.

But why did it take 50 years for the Nobel committee to decide that Gurdon's research was worthy of such an award? One reason was that Gurdon's results were not immediately and uncritically accepted by his colleagues, a number of whom suggested a variety of reasons that Gurdon might have drawn the wrong conclusions from his studies. For example, some critics thought that Gurdon's "mature" cells may have been contaminated with stray embryonic cells that had migrated into his sample. In such a case, Gurdon's experiments would have proved nothing more than researchers had already learned about the totipotency of embryonic cells. Briggs and King, themselves, raised such objections in the years following Gurdon's original papers on the cloning of animal cells (Crowe 2011; Kain 2009).

It took more than a decade for Gurdon's colleagues to fully accept the validity of his experimental results and to begin to appreciate their significance for future research on the cloning of animal cells. Beyond that point, two major issues in the field of animal cloning became more clear. First, researchers became increasingly interested in the challenge of cloning animals more complex than salamanders, newts, frogs, and the like; they began to think about the cloning of mice, rats, guinea pigs, dogs, cats, and, yes, perhaps humans. Second, it was obvious that increasingly more sophisticated techniques would be needed to carry out such experiments, although the precise nature of those techniques was certainly not fully understood.

A Timeline of Cloned Species

In the half century since Gurdon began his research on cloning, scientists have explored the possibility of using his technology for the cloning of a wide variety of animal species. The following timeline focuses on some of the most significant of those experiments.

1963

Anticipating the cloning research of Briggs, King, Gurdon, and others in the West on amphibians by at least two decades, Chinese embryologist Tong Dizhou (also called Tung Ti Chou) was exploring the possibility of cloning various species of fish. In the early 1940s, Tong carried out research on the cloning of blastomeres from goldfish much along the lines of later research on newts and salamanders. He came to the conclusion that this line of research could result in the cloning of fish and published two papers in 1944 and 1945 on his work. His most productive period of research was in the 1960s, when he conducted a number of successful cloning experiments with both goldfish and carp (especially *Rhodeus sinensis*). Tong's research was cut short by the Cultural Revolution in China and was never well known in the West because his publications were not translated from the Chinese (Zhu, Li, and Kang 2010; Zou 2013).

1986

Attempts to clone mice began as early as the mid-1970s, a not-surprising turn of events given the critical role that mice have always played in all forms of biological research. The first breakthrough in this line of research was announced in a paper published by Swiss biologist Karl Illmensee and American biologist Peter C. Hoppe in the January 1981 issue of *Cell*, a leading journal in the field (Illmensee and Hoppe 1981). Illmensee and Hoppe claimed to have cloned three mice using the technology developed by Briggs, King, and Gurdon. The journal was apparently sufficiently impressed that it placed a photograph of the three putative clones on the cover of that issue. The problem was that other researchers were unsuccessful in repeating Illmensee and Hoppe's research, and the two men reported that they too were unable to do so, for technical reasons. The problem became sufficiently troublesome that Illmensee's parent institution, the University of Geneva, instituted an investigation of the work that had been reported. That investigation concluded that no fraud had been committed, but Illmensee and Hoppe were not yet out of the woods. Soon after he was cleared of accusations of fraud by the university, Illmensee faced another set of charges from within his own department, questioning not only the cloning experiment itself but Illmensee's general conduct as a researcher. He eventually decided to leave the university and continue his work elsewhere. (He eventually became a staff researcher at the Andrology Institute of America.) Today, the Illmensee–Hoppe research is considered, at best, to be unconfirmed (Levine 2007, 50–52; MacKenzie 2009 [see link to original article on this affair]).

A more productive line of research with cattle was also reported in 1986 by a research team led by Danish embryologist Steen Willadsen. Willadsen produced clones of sheep by separating individual cells from an early-stage embryo. He then fused these cells with enucleated eggs with electrical impulses, producing "cloned embryos" that were then implanted in adult female sheep. The implanted embryos then grew and developed into normal lambs that were, of course, genetically identical to each other and to the parent donor sheep (Willadsen 1986).

1996

One of the most significant events in the history of cloning occurred in 1996 when Ian Wilmut, Keith Campbell, and their colleagues at the Roslin Institute of the University of Edinburgh, Scotland, and the biotechnology company PPL Therapeutics reported that a lamb that they named Dolly had been born as the result of nuclear transplantation technology. The significance of this research was, first, that it involved the cloning of a mammal, up to that time a relatively rare successful procedure and, second, the process involved the use of a somatic cell as the donor cell, rather than an embryonic cell, as had been the case in all previous such experiments. The cloning of Dolly then represented a repudiation of a very old tradition in cloning based on the belief that only the undifferentiated cells of an embryo could be used as donors in a transplantation experiment. Experiments such as the one that produced Dolly are now generally known as somatic cell nuclear transplantation (SCNT) procedures. Dolly later grew into a normal sheep who lived to the age of six, about a normal life span for an animal of her species. Questions have been raised as to whether she experienced medical problems not commonly seen in domestic sheep, although that issue has never been completely resolved (Dolly the Sheep 2013).

Post-Dolly

By the end of the 20th century, researchers had made significant advances in the science and art of cloning animals. One of the announcements made in 1999 concerned the birth of a Brahman bull calf cloned from the skin cells taken from a 21-year-old adult male just before he died. That bull, named Chance, had been somewhat of a celebrity, having been used in commercial photography and television shoots. His cloned son was given the obvious name of Second Chance and lived for eight years before dying of a digestive disorder that appeared to have been unrelated to the cloning experiment (Reunited [And It Feels So Good] 2005).

Two years later, another milestone in cloning was announced with the birth of a calf, who was eventually named Norm (for "normal"), to a two-year-old heifer named Daisy. The significant feature of the experiment was that Daisy had been cloned two years earlier from an older heifer, past menopause. Along with earlier experiments, such as the cloning of Chance in 1999, the experiment demonstrated the fact that SCNT using genetic material from old animals could still be used to produce healthy young animals and that there was no "sell-by date" that limited the use of such materials in cloning experiments (Bauman 2001).

Another breakthrough that occurred in 1999 was the cloning of the first male animal, a mouse that was given the name of Fibro. In an article in the journal *Nature Genetics*, University of Hawaii researchers Teruhiko Wakayama and Ryuzo Yanagimachi described how they removed the nucleus of cells taken from fibroblasts (hence the cloned mouse's name) from the tail of a common laboratory mouse and inserted them into the enucleated cells of female mice. The procedure resulted in the birth of an apparently normal clone of the original mouse, who later went on to sire two litters of pups (Abdulla 1999; Wakayama and Yanagimachi 1999).

Many of the advances in cloning research over the past two decades have raised new ethical questions about the practice. Prior to the 1990s, relatively few people were concerned about the possibility of cloning newts and salamanders, mice and rats, and even cows and sheep. But as researchers turned their attention to animals "closer to home," such as dogs and cats and humans, the debate over the ethical status of cloning spread far beyond scientific laboratories. Chapter 2 of this book discusses in more detail some of these questions about the ethics of cloning so-called higher animals. Some of the specific studies that include the following:

1999: Scientists at Oregon Regional Primate Research Center reported cloning the first nonhuman primate, a rhesus macaque they called Tetra (named for the use of tetraploid blastomeres in the cloning of the animal) (Chan et al. 2000).

2001: Researchers at Texas A&M University reported the first cloning of a domestic cat. The kitten born by the process was named CC (or C.C., for "carbon copy" or "copycat"). Five years later, CC had litter of kittens that were born normally and lived to adulthood (Berkowitz 2011; Braun 2002).

2001: The cloning of the first endangered animal was reported by the cloning firm of Advanced Cell Technology. The newborn gaur survived only two days before dying of a digestive disorder (Advanced Cell Technology, Inc. 2001; Appleton 2013).

2002: Researchers at the firm of Clonaid, headquartered in the Bahamas, report cloning the first human child. No independent confirmation of the report is ever produced, and the account is generally discounted by researchers in the field (Young 2002).

2003: A team of researchers at Spain's Center for Agro-Nutrition Research and Technology report that they have successfully cloned a bucardo, or Pyrenean ibex, the first officially extinct animal to have been cloned. The newborn survived only seven minutes, but researchers viewed its survival to birth as an important step in possibly restoring extinct animals to life on Earth (Choi 2009).

Another Approach to Cloning

One might simplistically describe the cloning experiments reviewed thus far as follows: The way to clone an organism is to extract the nucleus from a cell of that organism and transplant it into an enucleated cell. Since the transplanted nucleus contains all of the genetic information needed to form a new individual of that species, the procedure should result in a clone of the donor from whom the nucleus was taken.

Such a description is grossly over-simplified, of course. The problem is in the details, in discovering the precise techniques needed to produce a viable organism from the host egg. In reading reports of early experiments of this type, one is impressed by the rate of failure. Researchers were lucky if 5 percent of the host eggs produced by this procedure ever developed into a viable embryo. In the case of Dolly, for example, researchers transplanted donor nuclei into 277 host cells, of which 29 grew into embryos that could be transplanted into 13 surrogate ewes. Of those 29 embryos, only one grew to maturity, Dolly. Such numbers are typical of most experiments involving nuclear transplantation (Cloning Dolly the Sheep 2014).

Experiments of the kind discussed thus far are called *reproductive cloning* because their purpose is to produce offspring of an organism that is (or are) exact genetic copies of a parent organism. By the end of the 20th century, a second type of cloning experiment had become possible as a result of new information gained about the process of cellular reproduction and the development of new technologies for conducting research in this area. Many of these experiments fall into a category known as *gene cloning*, in which copies of a single gene are made.

The basis of gene cloning experiments is that scientists have long known that an organism's genetic information is stored in the nucleus of a cell. But until the middle of the 20th century, they knew almost nothing about the details of how the nucleus stored this information. It seemed obvious that some group of chemical compounds held the key to this puzzle, but just which compounds were they? Some scientists thought proteins were the most likely candidates for the storage of genetic information. They seemed the obvious candidates for this role because they are large, complex molecules capable of storing a great deal of very complex information. A minority of researchers put their faith in nucleic acids, also large molecules, but not nearly as complex as proteins. In 1944, however, Canadian-born American researcher Oswald Avery provided definitive proof that nucleic acids are the carriers of an organism's genetic information.

Oswald's discovery was a great breakthrough, but it was of somewhat limited value in the sense that neither Avery nor anyone else could say *how* nucleic acids store information. The answer to that question became available in 1953 when British chemist Francis Crick and American biologist James Watson discovered the three-dimensional structure of one type of nucleic acid, deoxyribonucleic acid (DNA). Some historians of science have ranked this discovery among the most important ever made in the history of science. How could they make such a judgment?

Prior to Crick and Watson's work, scientists knew some basic information about nucleic acids, specifically, the elements they contain (mostly carbon, nitrogen, oxygen, hydrogen, and phosphorus) and the relative abundance of each element in a nucleic acid. But they knew nothing about the way in which the atoms of these elements are connected to each other in a DNA molecule or what the general structure of the molecule looks like. It was like getting a box of Tinker Toys that listed on the outside of the box: 30 black Tinker Toys; 40 white Tinker Toys; 50 red Tinker Toys. That information would tell someone nothing at all as to how those Tinker Toys could be put together to make a house, a car, or a rocket. One would also need to know the shapes of the Tinker Toys and the number and shape of hooks, holes, and other connections. Crick and Watson provided all that information about DNA molecules.

The reason that the Crick–Watson discovery was so significant was that scientists now knew that DNA was a chemical compound, much like water, carbon dioxide, or sucrose (table sugar). It was a far more complex compound than any of those substances, but at least it was a chemical molecule whose exact structure could be determined and manipulated. A great deal of work would be involved, but scientists then had the capacity to handle and work with DNA molecules in exactly the same way they manipulated any number of other types of molecules. The next challenge for researchers was twofold: (1) what is the chemical mechanism by which DNA stores genetic information; and (2) what methods are available for manipulating DNA molecules, that is, for taking them apart, putting them back together again, changing their structure, and getting them to react with other substances?

The answer to the first question came in 1966 when research conducted by a number of workers over the preceding decade came to fruition with the discovery that DNA molecules use a genetic "code" that consists of the way in which three parts of the molecule, three *nitrogen bases*, are arranged along the spine of the molecule. (A three-base combination is known as a *codon*.) For example, a cluster of three nitrogen bases such as adenine–thymine–cytosine (abbreviated as A-T-C) coded for a specific amino acid, in this case, isoleucine. Overall, the 64 different three-base codes possible can code for all 20 of the amino acids used for making proteins in an organism (along with "start" and "stop" signals for use in making a protein). This breakthrough was also of enormous consequence, because it meant that, at least in theory, researchers could construct molecules that coded for any amino acid they wanted or even any combination of amino acids (i.e., proteins) they cared to make (Deciphering the Genetic Code 2014; a copy of the genetic code is included in this resource).

The answer to the second question was equally challenging, requiring the efforts of untold numbers of researchers over an additional decade of work. (Histories of scientific breakthroughs usually mention specific individuals who have led significant advances on a topic, for which they are often recognized with Nobel Prizes. Such a view is completely legitimate, although it tends to ignore the fact that dozens, hundreds, or thousands of other workers have contributed to those accomplishments.) One goal of this research was to find a way of obtaining or making a single gene and then learning how to manipulate that gene the way one might manipulate a molecule of carbon dioxide. If that goal could be accomplished, then scientists could go even further by combining genes in a variety of ways to produce the amino acids and proteins they wished to make.

A few basic technologies had to be developed to achieve this objective. One of the first steps was simply to isolate a single gene. Scientists had been talking about "genes" for more than a century, but no one had ever seen one or known (prior to the deciphering of the genetic code) what that term actually meant. In the summer of 1969, three workers at the Harvard Medical School, Jonathan Beckwith, Lawrence Eron, and James Shapiro, found a way to excise a single gene (called the lacZ gene) for the metabolism of lactose from the DNA of the bacterium *Escherichia coli* (*E. coli*). They used specially modified viruses to find a string of codons that codes for the lacZ gene; the viruses then attached the DNA molecule and snipped out the desired gene (Oberle 1969).

Two other basic technologies were needed if researchers were going to be able to manipulate molecules of DNA the way they work with other chemical molecules. They had to learn how (1) to cut DNA molecules apart and then (2) to put them back together again. These discoveries were actually made in reverse order, with the second step occurring in 1967 when five different laboratories almost simultaneously announced the discovery of molecules that are capable of joining two parts of a DNA molecule to each other, molecules known as *ligases* (Lehman 1974, 790; for an animated explanation of ligation, see DNA Ligation 2014).

The additional technology required for working with DNA involved the cutting apart of a molecule, a breakthrough announced in 1970 by a research team led by American microbiologists Hamilton O. Smith, Thomas Kelly, and Kent Wilcox. The "tool" discovered by Smith, Kelly, and Wilcox is called a *restriction endonuclease* or, more commonly, a *restriction enzyme*.

Restriction enzymes (REs) had actually been known since the early 1950s, although only in a form known as Type 1 RE. A RE is a molecule that cuts a much larger molecule, such as a protein or nucleic acid molecule, into two or more pieces so that the molecule can no longer function. Type 1 REs are not of much use in cloning experiments because researchers cannot control the changes they make in target molecules.

The REs discovered by Smith, Kelly, and Wilcox belong to a different category of REs, a category called Type 2 REs. Type 2 REs cut molecules only at very specific locations within the molecule, locations known as *recognition sites*. For example, the RE known as HhaI cuts a DNA molecule between a guanine nitrogen base and a cytosine nitrogen base in the sequence GCG*C, where * is the point of cleavage (Examples of Restriction Enzymes 2014).

At this point in history, scientists had all the tools they needed to carry out the process of gene cloning. The overall process involves the selection of a gene that codes for some desired characteristic and then finding a way to insert that gene into a host organism, often a bacterial cell. The cell is then cultured and allowed to reproduce multiple times. As the bacterial cell grows and reproduces, the inserted gene begins to function as it would in its normal setting. (A diagram of this process can be found at Gene Services 2010. The diagram shows a human cell, but the process is the same for all kinds of cells.)

Consider the process of cloning the human insulin gene. That gene, known as INS, codes for the production of insulin in the human body. Individuals who carry a mutated version of the INS gene are at risk for developing the disease known as diabetes mellitus. Those individuals can be treated for diabetes by receiving regular injections of the damaged INS gene. Here are the steps in producing synthetic insulin by gene cloning:

1. Researchers must find a copy of the normal INS gene. They can do so by one of two methods. First, the gene can be extracted from plants, animals, or other organisms that naturally contain the gene (a rather difficult and inefficient

process). Or, researchers can synthesize the gene in the laboratory. Recall that the gene for any characteristic is a chemical molecule that can be made artificially just as almost any other molecule whose structure is known can be made artificially.

2. The gene is inserted into some type of *vector*, a substance into which the gene can be inserted, and then inserted into a host cell. The most common type of vector used in gene cloning experiments is a plasmid, a circular piece of DNA found in bacteria and other unicellular organisms that is capable of replicating on its own. The process occurs when the plasmid is cut open by a restriction enzyme and the gene is annealed into the open plasmid.

3. The vector, now consisting of the plasmid with the inserted gene, is inserted into the host cell by any one of a number of means. This process is called *transformation*.

4. Once inserted into the host cell, the plasmids begin to reproduce along with the host cell itself. After the first cell cycle, two cells are formed, each carrying the transformed plasma; after a second cell cycle, four cells are formed; after a third cycle, eight cells are produced; and so on. In a relatively short period of time, the transformed host cell has produced hundreds or thousands of times. While propagation of the cell continues, the gene originally carried by the transformed vector continues to produce the substance for which it codes, in the example given here, insulin. This method provides, then, a method for producing very large amounts of insulin at a relatively low cost and in a relatively short period of time. (An animation of this process is available at Making Human Insulin 2014.)

The first production of human insulin by the process described here (also known as *genetic engineering*) was announced on September 6, 1978, by the Genetech company. The product, given the name of *Humulin*, was approved for human use four years later, on November 15, 1982. Today a number of genetically engineered forms of human insulin are available with a variety of properties, including Aspart, Glulisine, Humalog, Lantus, Levemir, Lispro, Novolin, and Tresiba. Gene cloning has also been used in the synthetic production of a number of other important products, as summarized in Table 1.2.

Table 1.2   Some Genetically Engineered Products Approved by the FDA

Product	Year Approved	Approved Use(s)*
Human insulin	1982	Diabetes mellitus
Human growth hormone	1985	Growth deficiency in children
Alpha interferon (2b)	1986	Hairy cell leukemia; genital warts; Kaposi's sarcoma; hepatitis B; hepatitis C
Hepatitis B vaccine	1986	Hepatitis B
Alteplase	1987	Acute myocardial infarction; acute massive pulmonary embolism
Epoetin alfa (erythropoietin)	1989	Anemia associated with chronic renal failure
Gamma interferon (1b)	1990	Chronic granulomatous disease; severe malignant osteoporosis
Antihemophiliac factor	1992	Hemophilia A
Proleukin (IL-2)	1992	Metastatic kidney cancer; metastatic melanoma
Beta interferon (1b)	1993	Certain types of multiple sclerosis
Imiglucerase	1994	Gauche's disease
Pegaspargase	1994	Acute lymphoblastic leukemia
Coagulation factor IX	1997	Factor IX deficiencies
Glucagon	1998	Hypoglycemia in people with diabetes
Adalimumab	2003	Rheumatoid arthritis
Platelet-derived growth factor BB	2005	Periodontal bone defects and associated gingival recession
Hyaluronidase	2005	Adjunct to injected drugs

(Continued)

Table 1.2 (Continued)

Product	Year Approved	Approved Use(s)*
HPV vaccine	2006	Human papillomavirus disease
Sorafenib	2007	Primary kidney cancer; primary liver cancer
Atryn	2009	Perioperative and peripartum thromboembolic events
Rixubis	2013	Prophylaxis and control of hemophilia B
Tretten	2013	Congenital factor VIII
Alprolix	2014	Hemophilia B

*Additional approvals may be given for similar or related products at a later date.

In addition to its use in the production of essential biological chemicals, such as insulin, gene cloning can be used in another way to improve the health of humans and other animals. Consider the case of diabetes again. People with this disease need to take insulin on a regular basis because their bodies either do not make enough of the substance themselves or they do not "know" how to use it properly. But suppose that there was a way to repair or replace the defective gene that causes diabetes in the first place. That might be a better solution to the problem of diabetes because the person with diabetes would not have to depend on an external source of insulin for the rest of his or her life.

Many researchers have high hopes that methods can be found to repair or replace the genes that cause diseases such as diabetes with gene cloning. The procedure used for this type of gene cloning, called *therapeutic cloning*, is essentially the same as the procedure described earlier for the production of insulin. Only the last step in the process differs because the modified gene produced by gene cloning is introduced into an animal (such as a human), where it then begins to function. In the case of diabetes, a correct form of the lacZ gene could be synthesized and multiple copies of the gene could be produced.

Those copies could then be introduced into a patient with diabetes, and, under the best of circumstances, the introduced gene could begin directing the production of insulin (Cloning Conundrum 2002).

Many people have had high hopes for therapeutic cloning as a means of curing genetic diseases of humans and other animals. But the process has been fraught with controversy and a host of technical problems, issues that will be discussed in greater detail in Chapter 2 of this book.

Conclusion

Progress in the field of cloning research has been remarkable over the past 150 years. Researchers have gone from an era when they knew how to produce many types of plant clones and a handful of animal clones to a day when they can synthesize virtually any type of clone that they can imagine.

Today, almost no one with the disease of diabetes mellitus needs to be concerned because adequate supplies of human insulin are available for their treatment.

But this chapter does not include some of even the most recent developments in this field, nor has it introduced any of the very contentious issues that surround the research on or possible applications of all types of cloning. Chapter 2 deals with all such issues.

References

Abdulla, Sara. 1999. "First Male Clone." *Nature*. http://www.nature.com/news/1999/990603/full/news990603-2.html. Accessed on November 28, 2014.

"Advanced Cell Technology, Inc. Announced That the First Cloned Endangered Animal Was Born at 7:30 p.m. on Monday, January 8, 2001." http://web.archive.org/web/20080501053203/; http://archive.is/QGuuZ. Accessed on November 29, 2014.

Amasino, Richard. 2005. "1955: Kinetin Arrives: The 50th Anniversary of a New Plant Hormone." *Plant Physiology.* 138(3): 1177–1184.

Anderson, Scott C. 2004. "A Baby's Hair." In Ann A. Kiessling and Scott C. Anderson, eds. *Human Embryonic Stem Cells.* Sudbury, MA: Jones and Bartlett, 2003. Available online at http://www.scienceforpeople.com/Essays/baby_hair.htm. Accessed on November 24, 2014.

"Animal Development: From Genes to Organisms: Experiment Links." 2014. http://bcs.whfreeman.com/the-lifewire8e/pages/bcs-main_body.asp?s=43000&n=00070&i=43070.01&v=chapter&o=|00030|&ns=0&uid=0&rau=0. Accessed on November 24, 2014.

Appleton, Caroline. 2013. "The First Successful Cloning of a Gaur (2000), by Advanced Cell Technology." https://embryo.asu.edu/pages/first-successful-cloning-gaur-2000-advanced-cell-technology. Accessed on November 29, 2014.

"Basics of Plant Tissue Culture." 2014. Value @ Amrita. http://amrita.vlab.co.in/?sub=3&brch=187&sim=1100&cnt=1. Accessed on November 18, 2014.

Bauman, David. 2001. "Clones from Aged Cows Have Normal Pregnancies and Calving." Uconn News. http://news.uconn.edu/2001/jun2001/rel01060.htm. Accessed on November 28, 2014.

Beetschen, Jean-Claude, and Jean-Louis Fischer. 2004. "Yves Delage (1854–1920) as a Forerunner of Modern Nuclear Transfer Experiments." *International Journal of Developmental Biology.* 48(7): 607–612.

Berkowitz, Lana. 2011. "First Clone Cat Turns 10." Chron. http://www.chron.com/life/article/First-cloned-cat-turns-10-1383844.php. Accessed on November 29, 2014.

Birnbaum, Kenneth D., and Alejandro Sánchez Alvarado. 2008. "Slicing across Kingdoms: Regeneration in Plants and Animals." *Cell.* 132(4): 697–710.

Braun, David. 2002. "Scientists Successfully Clone Cat." National Geographic News. http://news.nationalgeographic .com/news/2002/02/0214_021402copycat.html. Accessed on November 29, 2014.

Briggs, Robert, and Thomas J. King. 1952. "Transplantation of Living Nuclei from Blastula Cells into Enucleated Frogs' Eggs." *Proceedings of the National Academy of Sciences.* 38(5): 455–463.

Bryant, Geoff. 2006. *Plant Propagation A to Z: Growing Plants for Free.* Toronto: Firefly Books.

"Budding in Hydra." 2010. TutorVista.com. http://www .youtube.com/watch?v=a5oHMjGqjyo. Accessed on November 4, 2014.

Chan, A. W., et al. 2000. "Clonal Propagation of Primate Offspring by Embryo Splitting." *Science.* 287(5451): 317–319.

Choi, Charles Q. 2009. "First Extinct-Animal Clone Created." National Geographic News. http://news.nationalgeographic.com/news/2009/02/090210-bucardo-clone.html. Accessed on November 29, 2014.

"Clone and Characteristics of Clones." 2014. My Agricultural Information Bank. http://www.agriinfo.in/default.aspx? page=topic&superid=3&topicid=1793. Accessed on November 15, 2014.

"Cloning Conundrum." 2002. University of Wisconsin. http://whyfiles.org/148clone_clash/. Accessed on December 2, 2014.

"Cloning Dolly the Sheep." 2014. AnimalResearch.Info. http://www.animalresearch.info/en/medical-advances/ timeline/cloning-dolly-the-sheep/. Accessed on November 30, 2014.

Crowe, Nathan Paul. 2011. "A 'Fantastical' Experiment: Motivations, Practice, and Conflict in the History of Nuclear Transplantation, 1925–1970." http://conservancy. umn.edu/bitstream/handle/11299/119824/1/Crowe_ umn_0130E_12449.pdf. Accessed on November 26, 2014.

"Deciphering the Genetic Code." 2014. Office of NIH History. http://history.nih.gov/exhibits/nirenberg/index.htm. Accessed on December 1, 2014.

"Differences between Mitospores and Meiospores." 2014. Major Differences. http://www.majordifferences. com/2013/02/difference-between-mitospores-and.html#. VFpxQfnF-So. Accessed on November 5, 2014.

"DNA Ligation." 2014. New England BioLabs. https:// www.youtube.com/watch?v=Q3xVGvEGIsg. Accessed on December 1, 2014.

Dodds, John H., and Lorin W. Roberts. 1985. *Experiments in Plant Tissue Culture*, 2nd ed. Cambridge: Cambridge University Press. Available online at http://pdf.usaid.gov/pdf_docs/PNABD686.pdf. Accessed on November 22, 2014.

"Dolly the Sheep." 2013. Roslin. http://www.roslin.ed.ac.uk/ public-interest/dolly-the-sheep/a-life-of-dolly/. Accessed on November 27, 2014.

"Examples of Restriction Enzymes." 2014. Access Excellence. http://www.accessexcellence.org/AE/AEC/CC/re_chart .php. Accessed on December 1, 2014.

Fautin, Daphne. 2002. "Reproduction of Cnidaria." *Canadian Journal of Zoology*. 80: 1735–1754.

Ferguson, David J. P. 2009. "*Toxoplasma gondii*: 1908–2008, Homage to Nicolle, Manceaux and Splendore." *Memórias do Instituto Oswaldo Cruz*. 104(2): 133–148. Available online at http://www.scielo.br/scielo.php?script=sci_ar ttext&pid=S0074-02762009000200003. Accessed on November 4, 2014.

Gautheret, R. 1939. "Sur La Possibilité De Réaliser La Culture Indéfinie Des Tissues De Tubercules De Carotte." *Comptes Rendus des Séances de la Société de Biologie et de Ses Filiales*. 208: 118–120. (In French)

Gautheret, Roger J. 1983. "Plant Tissue Culture: A History." *The Botanical Magazine*. 96(4): 393–410.

"Gene Services." 2010. AbFrontier. http://www.abfrontier. com/cs/gene.do. Accessed on December 1, 2014.

George, Edwin F., M.A. Hall, and Geert-Jan de Klerk, eds. 2007. *Plant Propagation by Tissue Culture*. London: Springer. Available online at http://diyhpl.us/~bryan/ papers2/bio/agrobacterium/superkuh.com/library/ Biology/Agrobacterium/Plant%20Propagation%20 by%20Tissue%20Culture_%20%20Edwin%20 F%20George_%20Michael%20A%20Hall_%20 Geert-Jan%20De%20Klerk_%202007.pdf. Accessed on November 20, 2014.

Gilbert, Scott F. 2014. *Developmental Biology*. Sunderland, MA: Sinauer Associates, Inc. Publishers. A version of this text is available online at http://www.ncbi.nlm.nih.gov/ books/NBK9983/. Accessed on November 23, 2014.

"Grafting and Budding Nursery Crop Plants." 2014. North Carolina Cooperative Extension Service. http://www .ces.ncsu.edu/depts/hort/hil/ag396.html. Accessed on November 12, 2014.

Gurdon, John. 2009. "Nuclear Reprogramming in Eggs." *Nature Medicine*. 15: 1141–1144.

Gurdon, J.B., and J.A. Byrne. 2003. "The First Half-Century of Nuclear Transplantation." *Proceedings of the National Academy of Sciences of the United States of America*. 100(14): 8048–8052.

Harberg, Nicholas. 2011. "Why Plant 'Clones' Aren't Identical." Science Daily. http://www.sciencedaily.com/ releases/2011/08/110804212931.htm. Accessed on May 13, 2015.

Higgins, Adrian. 2005. "Why the Red Delicious No Longer Is: Decades of Makeovers Alter Apple to Its Core." The Washington Post. August 7, 2004. The Cabin Net. http:// thecabin.net/stories/080705/bus_0807050003.shtml. Accessed on November 11, 2014.

Highsmith, Raymond C. 1982. "Reproduction by Fragmentation in Corals." *Marine Ecology—Progress Series*. 7: 207–226.

"The History of Bartletts." 2014. Pears USA. http://web.archive.org/web/20070714194736/http://www.usapears.com/pears/varieties_yellow_bartlett.asp. Accessed on November 11, 2014.

"The History of Cloning." 2014. Learn.Genetics. http://learn.genetics.utah.edu/content/cloning/clonezone/. Accessed on November 24, 2014.

Hu, Ke, et al. 2002. "Daughter Cell Assembly in the Protozoan Parasite *Toxoplasma gondii*." *Molecular Biology of the Cell*. 13(2): 593–606.

Ikeuchi, Momoko, Keiko Sugimoto, and Akira Iwase. 2013. "Plant Callus: Mechanisms of Induction and Repression." *The Plant Cell*. 25(9): 3159–3173.

Illmensee, Karl, and Peter C. Hoppe. 1981. "Nuclear Transplantation in Mus musculus: Developmental Potential of Nuclei from Preimplantation Embryos." *Cell*. 23(1): 9–18.

Janick, Jules. 2005. "Horticultural Plant Breeding: Past Accomplishments, Future Directions." https://www.hort.purdue.edu/newcrop/pdfs/acta694.pdf. Accessed on November 11, 2014.

Janne, Everett E. 2014. "Air Layering for Difficult-to-Root Plants." Texas A&M AgriLife Extension. http://aggie-horticulture.tamu.edu/earthkind/landscape/air-layering/. Accessed on November 12, 2014.

Jiang, Caifu, et al. 2011. "Regenerant Arabidopsis Lineages Display a Distinct Genome-Wide Spectrum of Mutations Conferring Variant Phenotypes." *Current Biology*. 21(16): 1385–1390.

Jourdin, Sophie. 2010. "Philippe Van Tieghem (1839–1914) and the Approach of Plant Tissues Culture." *Annals of the History and Philosophy of Biology*. 15: 29–50.

Kain, Kristin. 2009. "The Birth of Cloning: An Interview with John Gurdon." *Disease Models and Mechanisms.* 2(1–2): 9–10.

Kavanagh, Frederick, and Annette Hervey. 1991. "William Jacob Robbins 1890–1978." *Biographical Memoirs of the National Academy of Sciences.* 60: 293–328.

Kearl, Megan. 2013a. "The Potency of the First Two Cleavage Cells in Echinoderm Development. Experimental Production of Partial and Double Formations" (1891–1892), by Hans Driesch. The Embryo Project Encyclopedia. http://embryo.asu.edu/pages/potency-first-two-cleavage-cells-echinoderm-development-experimental-production-partial-and. Accessed on November 23, 2014.

Kearl, Megan. 2013b. "Wilhelm Roux (1850–1924)." The Embryo Project Encyclopedia. http://embryo.asu.edu/pages/wilhelm-roux. Accessed on November 23, 2014.

Kraehmer, Hansjoerg, and Peter Baur. 2013. "Stolons and Runners." Chap. 49 in *Weed Anatomy.*

Krikorian, A. D., and David L. Berquam. 1969. "Plant Cell and Tissue Cultures: The Role of Haberlandt." *The Botanical Review.* 35(1): 59–67.

Krikorian, Abraham D., and Liisa Kaarina Simola. 1999. "Totipotency, Somatic Embryogenesis, and Harry Waris (1893–1973)." *Physiologia Plantarum.* 105(2): 347–354.

Kumar, G. N. M. 2014. "Propagation of Plants by Grafting and Budding." Pacific Northwest Extension Publication. http://www.coopext.colostate.edu/boulder/horticulture/pdf/Grafting%20Manual.pdf. Accessed on November 12, 2014.

Kumar, N. 2006. *Breeding of Horticultural Crops: Principles and Practices.* New Delhi: New India Publishing Agency.

Laimer, Margit, and Waltraud Rücker, eds. 2003. *Plant Tissue Culture: 100 Years since Gottlieb Haberlandt.* New York: Springer-Verlag.

Lehman, I. R. 1974. "DNA Ligase: Structure, Mechanism, and Function." *Science*. 186(4166): 790–797. Available online at http://www.sciencemag.org/content/186/4166/790. full.pdf. Accessed on December 1, 2014.

Lensch, M. William, and Christine L. Mummery. 2013. "From Stealing Fire to Cellular Reprogramming: A Scientific History Leading to the 2012 Nobel Prize." *Stem Cell Reports*. 1: 5–17.

Levine, Aaron D. 2007. *Cloning: A Beginner's Guide*. Oxford: Oneworld.

Maayan, Inbar. 2013. "John Bertrand Gurdon (1933–)." The Embryo Project Encyclopedia. http://embryo.asu. edu/pages/john-bertrand-gurdon-1933. Accessed on November 25, 2014.

MacKenzie, Deborah. 2009. "'Human Clones' Are Unlikely to Materialize." Short Sharp Science. http://www .newscientist.com/blogs/shortsharpscience/2009/04/ cloning-scientist-has-a-dark-p.html. Accessed on November 26, 2014.

"Making Human Insulin." 2014. ABPI Resources for Schools. http://www.abpischools.org.uk/page/modules/diabetes/ diabetes6.cfm?coSiteNavigation_allTopic=1. Accessed on December 1, 2014.

McKinnell, Robert Gilmore. 1985. *Cloning of Frogs, Mice, and Other Animals*. Minneapolis: University of Minnesota Press.

Melton, Douglas A. 2006. "Coaxing Embryonic Stem Cells." Potent Biology: Stem Cells, Cloning and Regeneration. http://media.hhmi.org/hl/06Lect3.html. Accessed on November 25, 2014.

Mudge, Ken, et al. 2009. "A History of Grafting." *Horticultural Reviews*. 35: 437–494.

Muir, W. H., A. C. Hildebrandt, and A. J. Riker. 1958. "The Preparation, Isolation, and Growth in Culture of Single

Cells from Higher Plants." *American Journal of Botany*. 45(8): 589–597.

Oberle, Mark W. 1969. "Harvard Team Isolates Gene." The Harvard Crimson. http://www.thecrimson.com/article/ 1969/11/24/harvard-team-isolates-the-gene-pa/. Accessed on December 1, 2014.

"Plant Propagation: Asexual Propagation." 1998. Arizona Master Gardener Manual. http://ag.arizona.edu/ pubs/garden/mg/propagation/asexual.html. Accessed on November 12, 2014.

Raghavam, V. 2003. "One Hundred Years of Zygotic Embryo Culture Investigations." *In Vitro Cellular & Development Biology—Plant*. 39(5): 437–442.

Rai, Rhitu. 2007. "Introduction to Plant Biotechnology." http://nsdl.niscair.res.in/jspui/bitstream/123456789/668/1/ revised%20introduction%20to%20plant%20biotcchnology. pdf. Accessed on November 13, 2014.

Reddy, Mallikarjuna, and Aparna Rao. 2010. *Plant Propagation in Horticulture*. New Delhi: Pacific Books International.

Reed, David W. 2007. "Cloning Plants: Cuttings and Layerings." Youth Adventure Program. http://hort201.tamu.edu/ YouthAdventureProgram/AsexualPropagation/AsexaulProp agation.html. Accessed on November 12, 2014.

Reif, Diane, and Elizabeth Ball. 2009. "Propagation by Cuttings, Layering and Division." Virginia Cooperative Extension. http://pubs.ext.vt.edu/426/426-002/426-002. html#L2. Accessed on November 12, 2014.

Reinert, J. 1959. "Über die Kontrolle der Morphogenese und die Induktion von Adventivembryonen an Gewebekulturen aus Karotten." *Planta: An International Journal of Plant Biology*. 53(4): 318–333.

"Resource Book on Horticulture Nursery Management." 2014. Yashwantrao Chavan Maharashtra Open University.

http://agropedialabs.iitk.ac.in/agrilore/sites/default/files/
HNM-book.pdf. Accessed on November 12, 2014.

"Reunited (And It Feels So Good)." 2005. This American
Life. http://www.thisamericanlife.org/radio-archives/
episode/291/transcript. Accessed on November 28, 2014.

Robbins, William J. 1922. "Cultivation of Excised Root Tips
and Stem Tips under Sterile Conditions." *Botanical Gazette*.
73(5): 376–390. Also available online at http://archive.org/
stream/jstor-2470362/2470362_djvu.txt. Accessed
on November 19, 2014.

Roux, Wilhelm. 1887. "Beiträge zur Entwickelungsmechanik
des Embryo: Über die Künstliche Hervorbringung Halber
Embryonen durch Zerstörung einer der Beiden Ersten
Furchungskugeln, sowie über die Nachentwickelung (Postgeneration) der Fehlenden Körperhälfte." ("Contributions
to the Development of the Embryo. On the Artificial
Production of One of the First Two Blastomeres, and the
Later Development (Postgeneration) of the Missing Half
of the Body.") *Archiv für Mikroskopische Anatomie*. 29(1):
157–212. Available online at http://link.springer.com/
article/10.1007%2FBF02955498. Accessed on November 23, 2014. (In German)

Roux, Wilhelm. 1883. "Über die Bedeutung der Kerntheilungs-
figuren." ("On the Significance of Mitotic Figures.") Leipzig:
Verlag von Wilhelm Engelmann. Available online at http://
babel.hathitrust.org/cgi/pt?id=coo.31924003154204;view=1
up;seq=7. Accessed on November 23, 2014. (In German)

Seabrook, John. 2011. "Crunch: Building a Better Apple."
The New Yorker. http://www.newyorker.com/maga-
zine/2011/11/21/crunch. Accessed on November 11, 2014.

Skoog, F., and C. O. Miller. 1957. "Chemical Regulation of
Growth and Organ Formation in Plant Tissues Cultured *In
Vitro*." *Symposia of the Society for Experimental Biology*. 11:
118–130.

Smith, E.A., et al. 2011. "Developmental Contributions to Phenotypic Variation in Functional Leaf Traits within Quaking Aspen Clones." *Tree Physiology*. 31(1): 68–77.

Spemann, Hans. 1902. "Entwickelungsphysiologische Studien am Triton-Ei." *Archiv für Entwicklungsmechanik der Organismen*. 15(3): 448–534. (In German)

Spemann, Hans. 1938. *Embryonic Development and Induction*. New Haven, CT: Yale University Press.

"Starfish (Seastars) Regenerating Their Arms with Tidepool Tim of Gulf of Maine Biological Supply." 2014. YouTube. http://www.youtube.com/watch?v=d5dOSyaKWTQ. Accessed on November 5, 2014.

Steward, F.C., Marion O. Mapes, and Joan Smith. 1958. "Growth and Organized Development of Cultured Cells. I. Growth and Division of Freely Suspended Cells." *American Journal of Botany*. 45(9): 705–708.

"Sumac Control." 2014. Pleasant Valley Conservancy. http://pleasantvalleyconservancy.org/sumac.html. Accessed on November 4, 2014.

Sunderland, Mary E. 2013. "Hans Adolf Eduard Driesch." The Embryo Project Encyclopedia. http://embryo.asu.edu/pages/hans-adolf-eduard-driesch. Accessed on November 13, 2014.

Sussex, Ian M. 2008. "The Scientific Roots of Modern Plant Biotechnology." *The Plant Cell*. 20(5): 1189–1198. Also available online at http://www.ncbi.nlm.nih.gov/pmc/articles/PMC2438469/pdf/tpc2001189.pdf. Accessed on November 18, 2014.

"T. Murashige and F. Skoog." 1962. https://essm.tamu.edu/media/46257/murashigeandskoogintropapersjanick.pdf. Accessed on November 20, 2014.

"T or Shield Budding." Texas A&M Agrilife Extension. http://aggie-horticulture.tamu.edu/earthkind/landscape/

plant-propagation/t-or-shield-budding/. Accessed on November 12, 2014.

Thorpe, Trevor. 2007. "History of Plant Tissue Culture." *Molecular Biotechnology*. 37(2): 169–180.

Thorpe, Trevor A. 2013. "History of Plant Cell Culture." In Roberta H. Smith, ed. *Plant Tissue Culture: Techniques and Experiments*. London: Academic Press.

Vasil, Indra K. 1985. *Cell Growth, Nutrition, Cytodifferentiation, and Cryopreservation*. Orlando, FL: Academic Press.

"Vegetative Propagation Techniques." 2007. Roots of Peace; USAID. http://www.sas.upenn.edu/~dailey/VegetativeProp pagationTechniques.pdf. Accessed on November 12, 2014.

Vrijenhoek, Robert C. 1998. "Parthenogenesis and Natural Clones." In Ernst Knobil and Jimmy D. Neill, eds. *Encyclopedia of Reproduction*. Vol. 3, pp. 695–702. San Diego: Academic Press.

Wakayama, Teruhiko, and Ryuzo Yanagimachi. 1999. "Cloning of Male Mice from Adult Tail-tip Cells." *Nature Genetics*. 22(2): 127–128.

"Walter Kotte: Wurzelmeristem in Gewebekultur." 1922. *Berichte der Deutschen Botanischen Gesellschaft*. 40(8): 269–272. (In German)

"The Weird and Wonderful Ways Plants Reproduce." 2014. http://dept.ca.uky.edu/PLS440/lectures/geophytes/Alterna tivepropagation.pdf. Accessed on November 4, 2014.

Weismann, August. 1893. *The Germ-Plasm. A Theory of Heredity*. New York: Charles Scribner's Sons. Available online at http://www.esp.org/books/weismann/germ-plasm/facsimile/.

"What Is Cloning?" 2013. Learn.Genetics. University of Utah. http://learn.genetics.utah.edu/content/cloning/whatiscloning/. Accessed on November 27, 2014.

White, Philip R. 1934. "Potentially Unlimited Growth of Excised Tomato Root Tips in a Liquid Medium." *Plant Physiology*. 9: 585–600. Available online at http://www.ncbi.nlm.nih.gov/pmc/articles/

PMC439083/pdf/plntphys00321-0188.pdf. Accessed on November 20, 2014.

White, Philip R. 1936. "Plant Tissue Cultures." *Botanical Review*. 2(8): 419–437. Available online at http://www.jstor.org/discover/10.2307/4353134?uid=3739960&uid=2&uid=4&uid=3739256&sid=21104594958171. Accessed on November 18, 2014.

White, Philip R. 1939. "Potentially Unlimited Growth of Excised Plant Callus in an Artificial Nutrient." *American Journal of Botany*. 26(2): 59–64. Available online at http://www.jstor.org/discover/10.2307/2436709?uid=3739960&uid=2&uid=4&uid=3739256&sid=21104594958171. Accessed on November 18, 2014.

White, P. R. 1951. "Nutritional Requirements of Isolated Plant Tissues and Organs." *Annual Review of Plant Physiology*. 2: 231–244.

Willadsen, S. M. 1986. "Nuclear Transplantation in Sheep Embryos." *Nature*. 320(6057): 63–65.

Wright, Cynthia. 1994/1995. "They Were Five: The Dionne Quintuplets Revisited." *Journal of Canadian Studies*. 29(4): 5–14.

Young, Emma. 2002. "First Cloned Baby 'Born on 26 December'." New Scientist. http://www.newscientist.com/article/dn3217-first-cloned-baby-born-on-26-december.html#.VHoGFTHF-So. Accessed on November 29, 2014.

Zhu, Zuoyan, Ming Li, and Le Kang. 2010. "Father of Biological Cloning in China." *Protein & Cell*. 1(9): 793–794.

Zou, Yawen. 2013. "Dizhou Tong (1902–1979)." The Embryo Project Encyclopedia. http://embryo.asu.edu/pages/dizhou-tong-1902–1979. Accessed on November 26, 2014.

On December 26, 2002, a company known as Clonaid announced the birth of the first human baby produced by cloning. The company later announced that four additional babies produced by their cloning process were also developing normally in utero and were expected to be born normally soon. No independent confirmation of these claims was ever produced, and many scientists doubt that the events announced had actually taken place. Much more to the point, however, was the fact that virtually all researchers realized that human cloning was no longer a fantastical idea; it had become, at the very least, a theoretical possibility. And with that realization has come a revolution in the way scientists and laypersons now think about the future of cloning. (Clonaid Claims Were a Hoax 2003; Young 2002)

While research on various types of cloning was going forward in the last decades of the 20th century, important breakthroughs in another area of biology were also occurring. These breakthroughs involved the study of stem cells. A stem cell is an undifferentiated cell with the capacity for developing into a differentiated cell. (There has been considerable debate about the precise way to define stem cells. For a review of this discussion, see Stem Cells: Scientific Progress and Future Research Directions 2001, 5–6.) The undifferentiated cells found in the

---

Dolly, the sheep, was the world's first cloned adult animal. She was born in 1996 at the Roslin Institute in Edinburgh, Scotland, and lived until February 14, 2003, when she died of natural causes. (AP Photo/PA/Files)

meristem of a plant that are capable of developing into specialized cells (plant stem cells), such as those found in roots, leaves, buds, and other plant parts, are examples of stem cells.

Developments in Stem Cell Research

The existence of stem cells in animals seems intuitively obvious. Any animal that begins life through the fertilization of an egg with sperm goes through early stages in which the fertilized egg divides into two cells, then four cells, then eight cells, and so on. Those cells are not specialized cells, capable of developing into muscle cells, another into skin cells, another into nerve cells, and so on. Instead, those initial cells must be capable of differentiating eventually into any one or another of the, for example, 210 to 240 different types of cells found in the human body.

The possibility that stem cells exist was hypothesized as early as 1868 by the German biologist Ernst Haeckel. Haeckel used the term *Stammzelle* to refer to a primitive cell from which all multicellular organisms developed. It was almost two decades later before American zoologist Edmund B. Wilson suggested the same term, *stem cell*, for those portions of a plant that involved in the growth of plants (Ramalho-Santos and Willenbring 2007, 35–36).

For more than a half century, many researchers continued to believe in the existence of stem cells, but they had little specific information on their properties and characteristics, their places of origin in animal bodies, the mechanisms by which they became differentiated, and other basic features. The first major breakthrough in this research logjam occurred in 1961 when James Till and Ernest McCulloch at the Ontario Cancer Institute demonstrated that stem cells really do exist. This groundbreaking discovery, for which Till and McCulloch are sometimes called the Fathers of Stem Cell Research, came about quite by accident. The two men were studying the effects of radiation on mice, with special attention to the factors that allowed mice to survive after receiving various doses of radiation. The experiment involved irradiating mice with dosages that would normally cause death within 30 days and then injecting those mice with various amounts of bone marrow cells to see how the amount of cells related to the probability of survival among the mice. On one particular day, McCulloch noticed the presence of a nodule on the spleen of one of the mice he was dissecting. He suspected that the nodule consisted of blood cells that had developed from a single stem cell contained in the bone marrow injection given to the mouse. He and Till concluded that the bone marrow cells they were using were actually clones of stem cells from donors with the capability of developing into mature differentiated cells that were keeping the mice alive (Maugh 2011; Till and McCulloch's original paper was Till and McCulloch 1961).

Till and McCulloch chose a rather obscure journal that was largely unknown to or ignored by their colleagues in the field. As a result, it took many years before other biologists learned about the discovery and were able to appreciate its significance to the field of stem cell research. As Till himself later said, there was a "long interval" of "at least 30 years" from the time he and McCulloch made their original discoveries until the biological community became really excited about the potential applications and benefits provided by stem cell research.

As evidence of that fact, it was exactly 20 years before the next major advance in stem cell research occurred. In 1981, University of California, San Francisco, researcher Gail Martin became the first person to isolate stem cells from a mouse embryo and demonstrate that they were, indeed, pluripotent stem cells. She accomplished this task by removing cells from a mouse embryo, culturing them in vitro, and then transplanting them into an adult mouse. She discovered that the host mouse developed a teratocarcinoma (teratoma), a mass of undifferentiated cells that later developed into differentiated nerve, muscle, blood, and other cells that would normally be produced during the development of a mouse embryo. Martin called the cells she removed from the mouse embryo *embryonic stem cells*, a term that has been retained to the present day.

Somewhat ironically, two researchers at the University of Cambridge, Martin Evans and Matthew Kaufman, reported similar results almost simultaneously with those of Martin. Evans and Kaufman reported that they had been able to remove stem cells from a mouse embryo, culture them in vitro, and then implant them into a live adult mouse, where the cells then developed normally in vivo as a teratoma, or allow them to continue developing normally in vitro (Evans and Kaufman 1981).

Once again, research on stem cells moved forward slowly for well over a decade before yet another major breakthrough occurred. By this time, researchers were pursuing the possibility of isolating and growing embryonic stem cells from more complex animals, such as monkeys and marmosets. The center for much of this research was the Stem Cell & Regenerative Medicine Center at the University of Wisconsin–Madison. Leader of most of the early research at the center was James A. Thomson, who continues to serve as director of regenerative biology at the University's Morgridge Institute for Research and professor of cell and regenerative biology, as well as professor of molecular, cellular, and developmental biology at the University of California, Santa Barbara. Thomson's research in the early 1990s involved the removal of embryonic cells from both Old World monkeys (the rhesus macaque) and New World monkeys (the common marmoset), the culturing of those cells in vitro, and their development and differentiation in essentially the same way as do normal embryonic cells.

The significance of Thomson's research was the possibility that similar experiments could be done with human embryonic stem cells, a possibility that was realized in 1998 both in Thomson's laboratory and in that of John D. Gearhart, and then at the Johns Hopkins University in Baltimore. Both research teams, using different research methods, isolated human embryonic stem cells, maintained them for an extended period of time *in vitro*, and then determined their characteristics and allowed them to develop through about 20 passages. The embryos were found to meet all criteria for being valid human embryos, but were eventually destroyed for ethical reasons (Stem Cells: Scientific Progress and Future Research Directions 2001, Ch. 3).

The Thomson and Gearhart research truly marked a turning point, not only in the field of stem cell research, but also in the wider field of cloning research. As one group of commentators wrote shortly after the Thomson and Gearhart discoveries:

> Readers will probably agree that few if any previous scientific papers reporting the charactcrisation of cultured cell lines would have attracted a similar degree of public attention. The interest stems in part from the ethical controversy surrounding the origins of the cells but chiefly from the widespread conviction that their availability will profoundly alter our approaches to many problems in human biology and medicine. (Pera, Reubinoff, and Trounson 2000, 5)

Pera, Reubinoff, and Trounson had put their finger on the reason that Thomson, Gearhart, and their colleagues had turned the world of cloning upside down. It was always one thing to use cloning to produce more appealing varieties of apples, to provide a new way of producing farm animals, and to devise new ways of producing valuable biological products. Such developments tended not to produce concern among the general public who, if they knew about it, probably had no objection to using science for the production of improved products in everyday life. But it was quite another matter to learn that that technology could also be used to clone human body parts or even whole and complete humans. Suddenly, the science of cloning became a matter of interest and concern not just to researchers in the field but also to politicians, ethicists, theologians, the medical profession, parents and loved ones dealing with life-threatening or debilitating diseases, and the general public.

Advantages of Plant Cloning

It is probably fair to say that the assessment of Pera, Reubinoff, and Trounson about the debate over cloning is of fairly recent vintage. As one looks back over the long history of plant cloning, for example, there is strong evidence that specialists in the field of horticulture and the general public at large had much to appreciate about plant cloning and very little, if anything, to object to. Some of the benefits of plant cloning that have been mentioned include the following.

Inefficiency of Natural Reproduction

One simple benefit of cloning plants is that Mother Nature is often not very efficient at reproducing plants on her own. A number of reasons exist for this fact. For example, naturally occurring plants may have defective genes that inhibit or reduce the process of reproduction so that a farmer or horticulturist can never be certain as to what the rate of reproduction of her plants will be in any given year. Rarely do all plants reproduce with complete efficiency, and a number of environmental and genetic factors may reduce the rate of reproduction to very low numbers. In addition, growers often develop plans with the specific goal of preventing their reproduction, by producing various types of seedless plants, for example. In today's world, consumers may have come to expect and demand only seedless fruits, such as seedless grapes and watermelons, which, by definition, are not able to reproduce naturally (because they have no seeds) (Burr and Burr 2002).

Efficiency of Cloning

By contrast, cloning can be used to produce large numbers of new plants from a small amount of parent plant material. Specialists in the field point out that thousands or even millions of new plants can be propagated from a single cutting through the use of tissue culturing. One writer has noted that it would not be unusual to produce more than a million chrysanthemum plantlets from a single cutting in one year (Reed 2007).

Tissue culturing also tends to produce more uniform and better-looking plants and fruit. Because the plantlets produced by cloning will almost always be exact copies of the parent plant, grower can be assured that the plants produced by tissue culturing will tend to have precisely the characteristics of the parent plant, the reason that the plant was chosen originally for use in the procedure. In instances where the producer needs to be sure that his or her product will be consistently appealing to the consumer, this property is especially important in plant reproduction (Goforth 2014).

Clones also tend to have a number of physical properties that make them more appealing to the consumer, whether that be the person purchasing a young ornamental plant for the front yard or the shopper who depends on the local grocery store to stock adequate supplies of attractive fruits and vegetables (Herren 2013, 171–174; Hershey 2002).

Cloning may also be the method of propagation of choice because mature plants develop more quickly, being able to bypass the planting, maturation, and development of the seed and seedling (Asexual Reproduction in Plants 2014). One goal of asexual propagation has been to reduce the period of juvenility of a plant, the time during which a plant has not yet reached reproductive age. In some plants, this period can be as much as 15 years, meaning that growers must wait a very long time before plants begin to bear fruit. One way around this problem is with some form of vegetative propagation, often grafting, a process that may cut the period of juvenility by at least half and as much as 75 percent (see, for example, Chong and Chai 1990).

Perpetuation of a Type of Plant

Cloning of a plant may also be necessary as the only economical method of perpetuating the survival of a particular desirable cultivar. Many types of spruce and maple trees, for example, grow much too slowly or not well enough by other methods of propagation and are therefore reproduced by grafting (Delius and Keegan 2014).

Elimination of Viruses and Other Plant Defects

Viral infections and other types of plant diseases are a serious problem for commercial plant growers who can easily lose a large fraction of their stock to such diseases. Although vegetative propagation can, on the one hand, be a mode of introducing viral diseases into clone plants, asexual reproduction is also a means for fighting such infections. One way to carry out the process is by regenerating plants that are known to be virus free by a variety of vegetative processes, ensuring that future generations of the stock will also be virus free (Applications of Cell and Tissue Culture 2014).

Better Commercial Products

For many plant growers, the bottom line for cloning plants is that the technology provides a way for dependably and reliably producing plants with desirable traits (color, taste, aroma, etc.) that reach maturity more quickly than if the plants are left to grow naturally, generally with a larger fruit output than is obtained from noncultured plants. In other words, cloning may allow the grower to produce a more desirable produce with greater profit than is possible with natural growing technologies (Cattle, McCreanor, and Mago 2008, slide 2; Higgins 2014).

Disadvantages of Plant Cloning

In spite of the many apparent benefits of plant cloning, some critics have long found reasons to oppose the technology, in part or in whole. One of the lingering concerns of some ethicists, theologians, scientists, and the general public is that researchers simply should not be changing the "natural order" of organisms by altering their fundamental genetic structure or the way by which organisms produce. They criticize those researchers for "acting against nature," trying to "play God," or otherwise disrupting the "natural order" of the universe (see, for example, Manninen 2014; Winston 2009, 61; Winter, Hickey, and Fletcher 1998, 359).

Among the more common concerns about plant cloning, however, is that such practices tend to eliminate genetic diversity, because all of the cloned plants in an area have almost precisely the same genetic composition. Genetic diversity is essential to the long-term survival of any population of plants (as well as animals) because it means that some plants will be more resistant to environmental risks than the average for the population. Suppose that a pest attacks a population of corn plants grown through sexual reproduction (the "natural" method of propagation). The pest may kill the vast majority of plants in the population, but simply because there is genetic variability in the population, a few plants will survive and the population can regrow. The same cannot be said for a population consisting of clones, because all plants have essentially the same genetic composition. If one plant is susceptible to attack by the pest, than all plants will be susceptible, and the pest will essentially wipe out the entire population of plants (Falk, Knapp, and Guerrant 2001). According to the Food and Agricultural Organization (FAO) of the United Nations, the tendency to clone plants specifically developed for desirable qualities has resulted in the loss of about 75 percent of the genetic diversity in plants grown commercially worldwide in the last one hundred years. As a result, humans now get about three quarters of all the plants they eat from only 12 different species (What Is Agrobiodiversity 2004; for more on the importance of genetic diversity see Consequences of Low Genetic Diversity 2014).

Related to issues of low genetic diversity is the potential problem of the inability of a cloned population to adapt to future environmental changes. Plants (and animals) must be able to deal with sudden attacks, such as those posed by pests, but they must also be able to adapt to long-term environmental changes, such as changes in climate or new variations in topographical or geographical features. Noncloned plants generally deal with such changes because, as with pests, they display genetic variability such that, even if a vast majority of a population is wiped out by long-term changes, some individual plants with unique genetic traits may be able to survive and allow the population to survive. (For a specific historical example of this situation, see Monoculture and the Irish Potato Famine 2014.)

It should be said that some cloned organisms are, in fact, able to change their genomic composition and evolve to adapt to new environmental conditions at least as efficiently as do non-cloned organisms (see, for example, Loxdale 2009).

A number of other possible issues have been suggested with the commercial use of cloned plants. In some cases, there is little strong evidence one way or another as to the legitimacy of the claim being made. In other cases, arguments can be made both for and against the claim.

- Shorter life span: Some writers suggest that cloned plants have a shorter life span than plants from the same species raised from seed. But there is also some evidence that cloned plants may also have a longer life—and can actually be engineered to have a longer life—than their native cousins (Cattle et al. 2008, slide 2; Life-Span 2014; Rooting Hormone for Plant Propagation 2014).

- Higher mutation rates: Although this topic has not yet been well studied, some writers suggest that cloned plants may have a higher rate of mutations than plants grown from seed. Depending on the nature of the mutations, the cloned plants might, therefore, be either more or less preferable as commercial crops (Cattle et al. 2008, slide 9; Jiang et al. 2011; Miyao et al. 2012).

- Higher costs: Experts in the field see the use of cloning for the propagation of plants as either a great benefit or a disadvantage, depending on a number of facts. At first glance, it is obvious that one can produce dozens, hundreds, or thousands of new plants with cloning technology compared to the much slower process of planting seeds for all such plants. On the other hand, cloning carries its own costs, including the expense of machinery and initial setups, the cost of hiring skilled workers, and the need to maintain sterile conditions in the workplace (Carroll 2014; Owen 2002).

- Perhaps the topic of debate of greatest interest with regard to the cloning of plants is its role in the production of genetically modified organisms (GMOs). A genetically modified organism is one in which the genome of a plant, animal, bacterium, virus, or other organism is altered by synthetic means. A new organism containing genes for a number of possible traits, such as resistance to certain pests, is then produced, grown, and reproduced by processes that involve cloning of either a gene or a complete plant. The debate over cloning, thus, becomes embedded in the larger question about the development, growth, and sale of GMO crops and foods. That debate is far too extensive and complex to be treated here, but a number of sources are available for further research on the controversy (see, for example, Newton 2014; Parsley 2004 [good art demonstrating the process]; Pighin 2003 [good art demonstrating the process]; Thompson 2015).

Some Applications of Cloning Procedures

As with the cloning of plants, all other types of cloning experiments and procedures are viewed as either positive or negative events by a variety of individuals and organizations. The next section of this chapter is devoted to a review of the risks and benefits, advantages and disadvantages, pros and cons, and other attitudes toward various types of cloning activities. Those activities are classified here into a handful of major categories: reproductive cloning; therapeutic cloning; cloning of endangered or extinct animals; and the use of stem cells for cloning.

Reproductive Cloning (Nonhuman)

Recall from Chapter 1 that the purpose of reproductive cloning is to produce offspring of an organism that is (or are) exact genetic copies of a parent organism. It is probably safe to say that the cloning of animals was of relatively modest concern to the general public until the birth of the first cloned sheep, Dolly, in 1996. At that point religious leaders, ethicists, scientific researchers, and ordinary men and women everywhere began to consider the ethical and moral aspects of cloning animals. Most of the arguments in favor of animal cloning were similar to those used for the cloning of plants, discussed earlier. For example, in a successful cloning operation, a person will know in advance almost exactly the type of animal that will be produced by the process; it will be virtually (but not entirely) identical to the animal from which it was cloned. As a result, someone who has an animal with especially favorable features—such as the ability to produce large quantities of milk, especially attractive hair or wool, resistance to disease, or a particularly appealing personality—has the ability to produce as many copies of that animal as he or she wishes (Hare 2003).

Just one example of the improvements that can be expected of cloning dairy animals is available from the laboratory of Kenneth White at Utah State University. White and his colleagues have been exploring the potential of cloning dairy animals for more than a decade and were responsible for the production of the world's first cloned mule, Idaho Gem, in 2003. White's research is based on a number of potential uses for cloned animals, one of which is their increased economic value. He reports, for example, that some of the Jersey cows that his group has cloned produce an average of 44,000 pounds of milk per year compared to an average of 18,000 to 25,000 pounds for noncloned Jerseys and Holsteins (Cattle Cloning 2014). The White group has also had significant success in cloning cattle that produce larger amounts of high-quality beef than noncloned animals (Cattle Cloning 2014; Cloning Research Continues at USU 2014).

The improvement of stock through cloning is certainly not without its critics. One of the current disputes concerns the use of cloning to produce high-quality racehorses. The logic is that breeders might want to collect DNA from horses that have been particularly successful during their lives and then clone a new generation of horses carrying that DNA. In such a case, a horse race might well involve a contest among a half dozen or more of the greatest race horses that have lived in the past few generations. Until 2013, such a scenario was unlikely, because all the major horse racing associations prohibited the use of cloned horses in their competitions. In July 2013, however, a jury in the U.S. District Court for the Northern District of Texas, Amarillo Division, decided that the American Quarter Horse Association's (AQHA) rules against the use of cloned horses in its races was illegal and that such horses could be entered in all AQHA events. If that ruling is upheld, the presumption is that cloned horses of all kinds will be permitted to participate in professional races (Cloning Lawsuit 2013; Finley 2013).

As the cloning of domestic animals becomes more wide-spread, one of the questions that has arisen is the safety of food products obtained from such animals. That is, many individuals have asked whether it is safe to drink the milk from a cloned cow, to eat the meat of a cloned cow, or to consume other products from cloned farm animals. Many people object to the sale and consumption of such products for a variety of reasons. They say that such foods may have been contaminated by the cloning process, that scientists don't know enough about the safety of such foods, that animals suffer unnecessarily because of the cloning process, that consumers are being misled by claims for the safety and nutritional value of such foods, and that, at the very least, such foods should be labeled to allow consumers to know that they are buying cloned foods and not their "natural" counterparts. For example, in 2014, the consumers' organization, Consumers Union (CU), submitted a petition to the U.S. Department of Agriculture asking that the agency require the labeling of all cloned foods so that consumers would know when they were buying such products. Comments from some of the 32,816 online signers of that petition reflect some of the views about cloned foods, such as the following:

- Cloning meat puts us all living things on this earth at risk!
- The safety of cloned products has not been adequately established.
- This is a danger to all people. Please don't allow cloning in our food.

- How about aligning with nature's laws and setting about making safe healthy conditions for food sources to be cultivated/raised in?
- It is totally wrong to use people as test subjects.
- Cloning animals is immoral.
- Cloning meat has been proven to increase animals' suffering.
- I don't want cloned meat nor milk; it's just gross. (Tell FDA You Don't Want Cloned Meat for Dinner! 2014; comments have been edited for spelling, grammatical, and typographical errors. The original CU petition can be found at Citizen Petition 2014. For a more detailed discussion of this issue, see Sharples 2008.)

To date, much of the debate over the labeling of food products from cloned animals is based on the fact that the U.S. Food and Drug Administration (FDA) has ruled that such foods are fundamentally identical to foods that come from traditional, noncloned sources and that they, therefore, require no special labeling or other marketing provisions. In its 2008 report, *Animal Cloning: A Risk Assessment*, the Center for Veterinary Medicine of the FDA concluded from a review of "hundreds of studies" that "edible products" from a wide variety of domestic animals "pose no additional risk(s) relative to corresponding products from contemporary conventional comparators." The animals included in this conclusion were healthy juvenile bovine clones, adult bovine clones, adult swine clones, adult goat clones, and the progeny of all of these animals. The only major category of domestic animals *not* included in this conclusion was sheep, for whom there was insufficient evidence (although authors of this report said that it was probably safe to conclude from studies of other species that sheep products were also probably safe for human consumption) (Animal Cloning 2008, 12–15; for a compendium of FDA documents on the safety of food products from cloned animals, see Animal Cloning 2014).

The FDA took this action in spite of the fact that public opinion polls suggest that food from cloned animals is not particularly popular in the United States. A number of polls (many of them taken in anticipation of the FDA's actions on the issue) suggest that about two-thirds of all Americans have unfavorable views of food produced from clone animals. In one study, for example, 60 percent of respondents said that they were opposed to the production of food from cloned animals, 41 percent said that they would not purchase such foods even if they were approved by the FDA, and a third said that drinking milk from cloned cows was "morally wrong." Similar results were obtained in a half-dozen other public opinion polls taken in the two-year period prior to the FDA's decision on food from cloned animals (Compilation and Analysis of Public Opinion Polls on Animal Cloning 2008).

Public opinion in the European Union (EU) tends to mirror that in the United States. In the latest Eurobarometer poll on the topic, about two-thirds of respondents said that they were "somewhat unlikely" or "not likely at all" to consume meat or milk produced from cloned animals (Europeans' Attitudes toward Animal Cloning 2008, 40). EU policy makers have tended to cease on results such as these to oppose not only the production of food from cloned animals in their own countries but also to prevent importation of such foods from the United States (EP Says No to Food from Cloned Animals 2014).

Another potential benefit of cloning animals is in the production of biological products, such as drugs and vaccines. The process involves the following major steps. First, a researcher removes cells from a cow, goat, or other animal that produces large amounts of milk. Second, the researcher inserts into those cells a gene for the production of some desired protein product, such as insulin. The nuclei from these altered cells are then inserted into the egg of the goat, cow, or other animal. When the animal produces milk, the milk also contains the protein for which the gene codes. Because the same procedure can be used with multiple animals, and because that gene is transmitted from generation to generation, the cloning procedure has produced an "animal farm" that generates very large quantities of the desired protein product over extended periods of time (Barrett 2010; Holguin 2002; Pohlmeier and Van Eenennaam 2008). The use of plants and animals for the production of biological products has become increasingly popular within the pharmaceutical industry and is generally known as *pharming*, based on the use of agricultural techniques to produce pharmaceuticals (Pharming for Farmaceuticals 2014).

Objections to animals with altered genetic makeup, so-called transgenic animals, have also been expressed. One concern is for the potential harm to animals used in such experiments. Critics point out to the very high rate of failure in gene transfer experiments with mammals and the likelihood of disruptive changes in animals born with foreign genes in their bodies. One example that is often mentioned involved the cloning of cows in New Zealand for the purpose of producing a human fertility drug in their milk. Three of the four calves used in the experiment died, however, before the age of six months because of problems with their ovaries. The ovaries grew to many times their normal size, about equal to that of a tennis ball compared to that of a marble (Deaths of Transgenic Calves at AgResearch's Ruakura Facility 2010; Gibson 2010; Pharming for Farmaceuticals 2014).

In many cases, the concerns expressed by critics of this procedure are somewhat vague, in the sense that there is still a great deal that researchers do not know about the cloning process and its effects on target animals. Because humans really don't know what the long-term effects of the procedure may be, they say, researchers should move slowly on the implementation of the new technology.

Of special concern in this area is that the cloning of mammals that began with Dolly may have been, critics say, the first step toward a much more serious procedure: the cloning of human beings. Indeed, some observers have suggested that the concern over the cloning of farm animals is really about the cloning of human beings, and not farm animals (Edge and Groves 2006, 323–324; King 1999). This argument often takes the form of a "slippery-slope" concern. The term *slippery slope* refers to the process of taking some action that everyone or most people agree is reasonable, safe, and perhaps useful. But the next step beyond that initial stage may be less reasonable, safe, and useful, and the step beyond that, even less reasonable, safe, and useful. In other words, to avoid some potentially serious or even disastrous event in the future, we should not take some action that seems harmless or innocuous *now*. So many critics who oppose the cloning of farm animals often explicitly or implicitly note that their real major concern is that such actions are really only the first step in the cloning of humans. (Of the many essays on this topic, see Devolder 2014; Smith 2005; Thompson and Harrub 2001. Among the many good articles about the benefits and risks of cloning nonhuman animals, see All about Animal Cloning 2010; Langwith 2012; Smith et al. 2000.)

Pet Cloning

Three specialized areas of the cloning of nonhuman animals are of special interest to many people: the cloning of pets, endangered species, and extinct species. The desire to clone one's pet is perhaps understandable to millions of people around the world. In many cases, a person can develop a relationship with a dog, cat, horse, or other pets that approaches that between human relatives. The death of a pet can be one of the most profound emotional experiences in a person's life. Under those circumstances, it is understandable that a person might be willing to spend very large amounts of money to "bring the pet back" in the form of a clone.

The history of pet cloning dates to 1998 when billionaire John Sperling, founder of the for-profit University of Phoenix, and his then wife Joan Hawthorne decided to attempt having their beloved border collie-husky mix, Missy, cloned. Sperling and Hawthorne felt that Missy was an extraordinary creature and that she should be replicated before her death, if that was possible. And given the birth of Dolly, the first cloned sheep, in 1996, Sperling and Hawthorne believed that the time might be ripe for such an event. To achieve this goal, Sperling established a project called Missyplicity, designed to produce a clone of the dog. The project was incorporated in 2000 in Sausalito, California, under the name of Genetic Savings & Clone (GSC) and produced its first cloned pet, a cat named CC (for Copy Cat or cloned cat) in 2001. Unfortunately, Missy herself died in 2002 before a successful cloning procedure could be completed. GSC was somewhat successful in cloning cats and sold the first commercially cloned cat, Little Nicky, to a private party for $50,000 in Texas. The company never achieved much success in producing a cloned dog, however, and it went out of business in 2006. (For the last available Web site presentation for the company, see Genetics Savings & Clone, Inc. 2014.)

The campaign to produce cloned pets did not die with the dissolution of GSC, however. In 2006, a group of South Korean researchers created the Sooam Biotech Research Foundation with the objective of developing the technology for cloning dogs, cats, and other pets, as well as the cloning of other animals for research purposes. The chief technology officer at Sooam, Researcher Hwang Woo-Suk, is otherwise best known for being implicated in the falsification of reports on the first cloning of a human embryo in 2005 (Cyranoski 2014). Sooam has announced the successful cloning of a number of animals since its founding, including Lancelot, the world's first commercial dog clone; Trakr, who worked in recovery efforts at the 9/11 terrorist attack; transgenic dogs for the study of diabetes and Alzheimer's disease; and a Chinese champion Tibetan mastiff (Dean 2014; History 2014). Sooam claims to have cloned more than 500 dogs in its history and now conducts raffles to determine who will have an opportunity to have her or his dog cloned next at no cost (Kutner 2014).

As appealing as cloning may be to some pet owners, critics have raised a number of objections to the practice. The American Anti-Vivisection Society (AAVS), for example, has produced a now-somewhat-outdated pamphlet discussing some of the major reasons that pet cloning is not a good idea. The

organization argues, to begin with, that cloning is a very inefficient and inhumane procedure. A successful clone may be produced only once in a few dozen, few hundred, or few thousand
attempts, and clones that are born may suffer from medical and
health problems that cause the animals pain and discomfort
and/or shorten their lives. The AAVS document quotes from
a number of authorities in the field, including one report that
notes that

> In all mammalian species where cloning has been
> successful, at best a few percent of nuclear transfer embryos
> develop to term, and of those, many die shortly after
> birth. . . . Even apparently healthy survivors may suffer
> from immune dysfunction or kidney or brain
> malformation, perhaps contributing to their death at later
> stages. Most frequently cloned animals that have survived
> to term are overgrown, a condition referred to as "large
> offspring syndrome." (Pet Cloning: Separating Facts from
> Fluff 2005, quoting Rideout, Eggan, and Jaenisch 2001)

Another problem with cloning pets is that the cloned puppy
or kitten may carry more than 99 percent of the same genes
as its parents, and even look a lot like them, but neither of
these facts means that the new animal will be very similar to
its parents. As an animal grows, both its physical and emotional characteristics begin to diverge from those of its parents
until, at adulthood, it may seem very different from them. In
its 2003 report on CC, the cloned cat, for example, *USA Today*
listed the ways in which the now-grown kitten differed from
its mother, one being slim and the other chunky; one being
adventurous and playful and the other shy and reserved (1 Year
Later 2003; Matulef 2013; for a cautionary tale in this regard,
see Morgenstern 2011).

Many critics also point to the considerable expense involved
in cloning a pet. Prices quoted by Sooam and other potential
cloning agencies run as high as $100,000 or more per cloning

event (Back from the Dead 2012; Cloning Your Dog 2014), an expense that some observers have called absurd, preposterous, vain, or indefensible in light of other ways in which such large amounts of money can be spent. (See, for example, 150K to Clone a Pet? 2014.) Rudolf Jaenisch, an expert in the area of cloning, professor of biology at the Massachusetts Institute of Technology, and founding member of the Whitehead Institute, has called the cloning of pets "ridiculous" and "preposterous" (Goldenberg and Jha 2004; Quick 2005, 88).

The bottom line in this debate may well be that, in the absence of legislation that forbids the cloning of pets, some individuals will feel that it is well worth whatever expense is involved to have a clone of their dog, cat, or other pets. Many Internet blogs contain comments that illustrate the point that people who have—or might have—the money needed to clone a pet would not hesitate to do so. (See, for example, Woman Pays $50,000 to Clone Dog 2012. An excellent general reference on the cloning of pets is Woestendiek 2010.)

Cloning of Endangered Species

Another area in which cloning has drawn some attention involves the cloning of endangered species. Endangered species are species that are regarded as being in imminent danger of extinction by authorities in the field. Some species are also regarded as *threatened* if they face serious threats to their survival that are not as great as those endangered species face. Wildlife biologists and other concerned specialists have looked for a variety of ways of protecting endangered species, such as provision of protected areas for such animals, breeding programs in zoos and other protected facilities, and removal of threats (e.g., pesticides and lead gunshot) to their survival. A few biologists believe that cloning endangered species may provide a way of ensuring that at least some individuals in a species will survive, even if only on a short-term basis while other programs for survival are being developed. (The process of trying to save an endangered species is sometimes referred to as *de-extinction*.)

Probably the primary reason to consider the cloning of an endangered species is that it might be the only viable method for keeping the species from becoming extinct. Most biologists would probably prefer to find ways to protect endangered species in their natural environment, but sometimes that option is simply not available. In December 2014, for example, one of the five last remaining white rhinoceroses in the world died in the San Diego Zoo at the age of 44. At that point, there no longer seemed any way to keep the species from going extinction in the next few years. In such cases, cloning offers at least a theoretical possibility of bringing a species back from the brink of extinction.

A number of attempts to achieve this objective have already been made. The first success was achieved in 2000 when researchers at the U.S. cloning company, Advanced Cell Technology, announced that they had cloned a gaur (*Bos gaurus*), an Asian ox then listed by the International Union for the Conservation of Nature as endangered (Bos gaurus 2014). A single animal, which researchers named *Noah*, was born in an experiment that began with 692 enucleated eggs, of which 81 grew to the blastocyst stage and about half of which were implanted into 32 cows. Of the 32 cows, 8 became pregnant, half of whom aborted spontaneously. Only Noah was born, and he survived only two days before dying of dysentery (Appleton 2013; Lanza, Dresser, and Damiani 2000).

Other attempts to clone endangered species have involved the mouflon (*Ovis orientalis musimon*) (Loi et al. 2001), banteng (*Bos javanicus*) (Nijman et al. 2003), African Wildcat (*Felis silvestris lybica*) (Gómez et al. 2004), and wild coyote (*Canis latrans*) (Hwang et al. 2013).

These experiments have only encouraged the debate over the legitimacy and value of cloning endangered species. On the one hand, supporters of the technology point out that it is often the only possible way to keep a species from going extinct, that cloned animals can act as breeding stock from which the species might be regenerated, and that, in any case, little or no harm can be done by making the effort to save a species. On the other hand, critics of the process claim that efforts should be devoted to other means of saving a species, such as habitat preservation; that the clones produced will not be healthy breeding stock because of the inbreeding involved in the reproduction of the species; and that the cost of the process is so great and the chances of success are so low that cloning can never become a practical means of saving any species (Brand 2014; Friese 2013; Jabr 2013).

One of the most ambitious programs for the cloning of wild animals is one scheduled by conservation groups in Brazil. The Brasilia Zoological Garden is working with the nation's agricultural research agency, EMBRAPA, to clone eight of the country's most endangered animals: the maned wolf (*Chrysocyon brachyurus*), jaguar (*Panthera onca*), black lion tamarin (*Leontopithecus chrysopygus*), bush dog (*Speothos venaticus*), coati (*Nasua*), collared anteater (*Tamandua tetradactyla*), gray brocket deer (*Mazama gouazoubira*), and bison (*Bison*). As of early 2015, the project is still in its early stages, which involve the collection of blood, sperm, and other genetic material for use in cloning, along with the training of personnel to carry out the necessary procedures. There is currently no plan to release clones into the wild, but to make sure they are available if the status of a species becomes more problematic in the future (Marcondes 2012).

Cloning of Extinct Species

The term *de-extinction*, used earlier, has another meaning in addition to its being an attempt to keep endangered species from going extinct. It also refers to the possibility of creating animals that have already gone extinct, such as the passenger pigeon or dodo bird. From one standpoint, there wouldn't seem to be much point in trying to recreate an extinct species, which, in most cases, went extinct because it could not survive in changing environmental conditions. Also, the idea of producing enough new members of an extinct species to create a viable population seems remote. And if that goal could be achieved, where would one place the extinct species so that it could once more flourish as a viable population of animals? In addition, biological purists doubt that one could propagate an animal that is truly a member of an extinct species because of differences in genomes between the original species and the new individual. Perhaps the cloning of an extinct species would be a fascinating scientific experiment, but is it likely to be of any general interest at any time in the future? (For comments on the inadvisability of cloning extinct species, see, for example, Brand 2014; Mark 2013; Pimm 2013; Shanks 2013.)

Proponents of cloning extinct species have a very different view of the technology. One reason for bringing back extinct species, they say, is that the return of such animals could provide researchers with a wealth of information about species that once thrived on Earth, even though they have long disappeared. It is also possible that the renewal of a species could contribute to the protection, renewal, or survival of various ecosystems in which the species once lived. Proponents also raise the question of justice with respect to the fact that humans were responsible for the elimination of untold numbers of species, so perhaps it is only fair if they now make possible the revival of some of those species. Finally, supporters of de-extinction often mention the "wow" factor, pointing out how "cool" it would be simply to see an animal whose species had disappeared from the face of the Earth a hundred or a thousand years ago. (For comments in support of de-extinction, see Brand 2013; Poinar 2013; Sherkow and Greely 2013.)

Whatever the opinion of scientists and the general public, the cloning of extinct animals has already occurred, and research in that direction continues today. The first cloned extinct animal was a bucardo, also known as a Pyrenean ibex (*Capra pyrenaica pyrenaica*). The last living example of the species was a free-ranging animal named Celia, who lived in the Monte Perdido National Park of Spain. In 1999, researchers decided to make an effort to save the species from extinction by taking samples from Celia to be used in cloning her. The project had just begun when Celia was found dead a year later, but it continued anyway. In 2003, a kid was born from the research, but it lasted only about seven minutes, dying as the result of a respiratory disorder (Folch 2009; Zimmer 2013). Since this event, research has continued in an effort to de-extinct a number of other species, including the thylacine (Tasmanian tiger—*Thylacinus cynocephalus*; Sanderson 2008); gastric-brooding frog (*Rheobatrachus*; Quirós 2007); and auroch (*Bos primigenius*; Mark 2013; Quirós 2007; Zimmer 2013).

Human Reproductive Cloning

Almost certainly, the cloning issue of greatest concern to most scientists and the general public is the cloning of human beings. Within days of the announcement of the cloning of Dolly at the Wilmut Institute in Scotland, President Bill Clinton announced that he was issuing an executive order suspending the use of federal funds for research on the cloning of human beings in the United States. In writing to his National Bioethics Advisory Committee (NBAC), Clinton said that

> While this technological advance could offer potential benefits in such areas as medical research and agriculture, it also raises serious ethical questions, particularly with respect to the possible use of this technology to clone human embryos. (Cloning Human Beings 1997, unnumbered page at introduction)

He therefore asked the commission to study this question and issue a report on its findings. (For the commission's final report, see Cloning Human Begins 1997.) Clinton received widespread support for his request for a voluntary moratorium on human cloning research (although very little support for legislation that *required* a cessation of such work) from the scientific community. (For details on this response, see Rossant 1997.)

The discussion about the pros and cons of the cloning of human beings has been conducted for nearly two decades in a variety of print and electronic media (see the end of this section for some suggested resources). Arguably one of the best summaries of that discussion can be found in a paper written for the NBAC in 1997 by Dan W. Brock, who was the then director of the Center for Biomedical Ethics at Brown University. Brock divides his conclusions into a few major categories, including the moral, individual, and social benefits and harms that might result from the cloning of humans. He begins with a long and relatively sophisticated discussion of the morality of human cloning; that is, he inquires as to whether humans have some fundamental "right" to be cloned and/or whether society at large has reasons to prohibit humans from making those individual choices. He concludes that the answers to these questions are "yes" and "no"; that is, humans do have a fundamental moral right to reproductive freedom, which involves the right to be involved in a cloning procedure, and cloning poses no moral threat to society at large, so it can impose no restrictions on a person's involvement in the cloning process (Brock 1997, E4–E6 and E-11–E-14; for similar arguments see Appel 2009; Breitowitz 2002; Shikai 2004).

As to be expected, a number of writers have taken a very different view about the morality and ethics of human reproductive cloning. For example, in a staff working paper prepared for the February 13, 2002, meeting of the President's Council on Bioethics, the argument is made that reproductive cloning of humans is not a morally acceptable procedure because "human manufacture [of clones], guided by market principles, violates some fundamental principles of human dignity and moral conduct; and "reproductive" cloning could make such violations easier and thus more common" (Staff Working Paper 3B 2002). One of the fundamental principles of morality that might be violated, according to some observers, is the right of every individual to complete self-determination. A person who is cloned is inevitably assigned his or her parent's genome, and the right to self-determination is therefore violated. Christof Tannert, the then chair of the Research Group for Bioethics and Science Communication at the Max Delbrück Centre for Molecular Medicine in Berlin, Germany, put the argument in these terms:

> Even if no one can be autonomous in the stipulation of his or her own genome, and even if natural procreation can, at least, include egoistic motives for the desired child, the arbitrary production of a genetically identical person is ethically reprehensible because the egoism of the clone generator restricts the clone's individuality. (Tannert 2006; also see Devolder 2014; Manninen 2014; Reproductive Cloning 2004; Reproductive Cloning: Ethical and Social Issues 2004)

Brock then goes on to review a number of possible individual and social benefits that might accrue from the availability of human reproductive cloning to the general public, including the following (all quotations are from Brock 1997):

- "Human cloning would be a new means to relieve the infertility some persons now experience." Many married couples are not able to have children because they are unable to produce viable or sufficiently healthy eggs or sperm. They may also be susceptible to other disorders of the reproductive system that prevent normal pregnancies. Reproductive cloning would provide a way for such individuals to have children who are truly their genetic offspring rather than having to rely on a variety of other artificial reproductive technologies, such as in vitro fertilization or surrogate parenthood. Same-sex couples could also rely on reproductive cloning to have genetically related children (Devolder 2014; Doig 2003; Orentilcher 2001).

- "Human cloning would enable couples in which one party risks transmitting a serious hereditary disease, a serious risk of disease, or an otherwise harmful condition to an off-spring, to reproduce without doing so." Scientists know of many physical disorders that are caused by genetic errors, including such life-threatening conditions as cystic fibrosis, muscular dystrophy, Tay-Sachs disease, hemophilia, phenyl-ketonuria, and various forms of cancer. Many couples choose not to have children because of the risk of transmitting one of these diseases to their offspring. In many cases, reproductive cloning may offer a possible option for such couples to have children who are truly their genetic offspring (Gillam 2008; Robertson 1994).

- "Human cloning of a later twin would enable a person to obtain needed organs or tissues for transplantation." The possibility has been suggested that scientists could create cloned twins so that one child would be available to provide "spare parts" for the other in case the other child was born with or has developed a life-threatening disorder. In such a case, the cloned twin could be sacrificed to provide the organ(s), tissue, or cells needed to allow his or her twin to survive. The technology for such a process would probably be available, but the ethics of carrying out such a procedure are profoundly difficult. How could one justify bringing a clone child into the world knowing that the child's primary purpose was to serve as an "organ factory" for his or her twin, some critics ask? A practical question might also be whether or not one would actually produce twin clones who would be born and raised at the same time or whether one would use the procedure only if a serious disease developed in a child or adult (Conger 2014; Dixon 2014). This option is not entirely a hypothetical issue as the cloning of a child for "replacement parts" has been attempted—and achieved—at least once in recent history (Inbar 2011; Savulescu 1999, 92).

- "Human cloning would enable individuals to clone someone who had special meaning to them, such as a child who had died." In a bizarre sort of way, this argument reflects the earlier discussion about the use of cloning to replace dead or dying pets, except for the fact that the argument can be made even more strongly about the loss of a child, a spouse, or other family member. The principles of reproductive cloning certainly make possible the cloning of the dead or dying family member, but the question is whether or not the procedure is justified for such a purpose. Some observers believe that however society as a whole may feel about such issues, it is still the right of an individual to make up his or her mind as to whether to use cloning for this purpose. But opponents of the practice raise a number of objectives to its use for this purpose. For example, they point out that a cloned child or spouse will never be precisely like the lost individual (as was pointed out for the cloning of pets earlier). Also, any cloned child might grow up with psychological issues that arise out of knowing that her or his main reason for life was to replace a progenitor. Also, the idea of using technology to produce a "replacement" individual contains a certain amount of the so-called yuk factor that often arises in discussing reproductive cloning, the feeling that such a procedure "just isn't right" (Devolder 2014; Kass, 2002, Ch. 5; Manninen 2014).

- "Human cloning would enable the duplication of individuals of great talent, genius, character, or other exemplary qualities." The idea that cloning could be used to produce individuals with extraordinary talents in one field or another is hardly a new one. In his 1963 essay, "Biological Possibilities for the Human Species in the Next Ten Thousand Years," the renowned British evolutionary biologist and futurist J. B. S. Haldane opined that cloning could be used to produce "exceptional people" such as great dancers, athletes, and poets. He saw cloning as perhaps being particularly useful for helping such extraordinary individuals avoid unhappy childhoods that are often the fate of boys and girls who stand out from the crowd and are subjected to bully-ing and censure by their peers. Haldane also saw cloning as a way of producing individuals with other types of special advantages, such as those with "permanent dark adaptation, lack of the pain sense, and special capacities for visceral per-ception and control" (Haldane 1963). The problems associ-ated with such a practice are probably well known and reflect many of the concerns about the once-popular "science" of eugenics, in which governmental bodies decided which in-dividuals or groups of individuals were "better" than others and deserved to be preserved and continued. Brock himself reviews many of the very difficult moral and ethical issues of actually putting into practice ideas such as those offered by Haldane (Brock 1997, E-9–E-10). In spite of these dif-ficulties, however, some observers continue to see cloning as a viable and desirable way of picking and choosing the types of individuals with special qualities who might be born in the future (see, for example, Volney 2013).

- "Human cloning and research on human cloning might make possible important advances in scientific knowledge, for example about human development." Somewhat ironi-cally, Brock wrote his essay about reproductive cloning just prior to the public announcement of the success by Thomson and Gearhart in the isolation and culturing of human stem cells. (See Chapter 1, pages 60–61.) His analysis of the use of reproductive cloning for research on human development was, therefore, somewhat less trenchant and extensive as it might have been. Since Thomson and Gearhart's discover-ies, of course, scientists have envisioned a very large array of ways in which cloned human cells can be used for studies on human embryogenesis and later stages of development. Most of those projects, however, involve the production of human embryonic stem cells, which are then destroyed at a later stage before they have undergone very much development. The creation and destruction of such cells has become the basis for a very extensive and often vitriolic debate between researchers and those who oppose such a practice on moral and ethical grounds. The details of that debate go far beyond the space available in this book, but it is safe to say that the idea of cloning human cells primarily for the purpose of carrying on medical research is still a topic of vigorous debate within the scientific community, as well as among the general public (Devolder 2015; Furcht and Hoffman 2011; National Research Council et al. 2010; Perrin and King 2014).

- "Human cloning would produce psychological distress and harm in the later twin." Brock and many other commentators have suggested that this risk of reproductive cloning might be the most serious of all such problems. They note that a child born through cloning might be under the impression, correctly or not, that her or his future is already determined, laid out by the genetic features with which she or he was born. Experts might point out that environmental factors are very important in the future one faces, but such assurances might not be sufficient to significantly impact one's view of one's own life to come. These commentators argue that the loss of a sense of personal freedom and choice in one's life might produce lasting psychological harm to the cloned child. A key element in this debate is often a concept developed by philosopher Joel Feinberg who thought that children were born with a fundamental right to foresee a variety of possible futures from which to choose for himself or herself, that is, to choose the kind of person he or she wants to be (Feinberg 1980). Again, whether correct or not, the cloned child may never see his or her future in that light, so that cloning becomes a deadly psychological influence on that person's life (Mameli 2007).

- "Human cloning procedures would carry unacceptable risks to the clone." Brock wrote his analysis of the risks and benefits of reproductive cloning during a period when the procedure was still very new. Statistics about the fraction of healthy live births of animals conceived by cloning technology were very discouraging. He argued that the existing data suggested that the chances of a healthy live child being born through reproductive cloning was very low and that such a child would far more likely be born with genetic defects that might presage a life of illness and perhaps an early and unpleasant death (Brock 1997, E-16–E-17). In point of fact, that outlook has not changed very much over the two decades since Brock offered his analysis. Current writers still emphasize the potential health threats to animal clones produced by reproductive cloning, such as cows and sheep. And although no human clone has ever been produced, it seems fair to say that any such event resulting from modern technology would also face very low chances of success (Devolder 2015). (Also see the position statement on reproductive cloning at American Association for the Advancement of Science Statement on Human Cloning 2013.)

- "Human cloning would lessen the worth of individuals and diminish respect for human life." Reproductive cloning may not only cause individuals to devalue their own lives, Brock suggests, but it may also cause society as a whole to think less highly of human life. The further that actual procreation wanders from the "natural" process that occurs between a man and a human, according to this position, the less "human" that activity may seem. This view is one that is held and advanced by specialists from a variety of fields, ranging from science and technology to philosophy and theology. For example, a group of eminent physicians, including the then surgeon general C. Everett Koop, wrote a paper in 2003 explaining why they thought that reproductive cloning of humans should not be permitted. Among the reasons they offered for their position was that

> "whether cloning were to become a widespread or an occasional practice, its acceptance would shift societal attitudes away from appreciating people as distinct

individuals and toward a new way of sizing up people as useful or attractive commodities of technology assembled to satisfy others' expectations." (Cheshire et al. 2003, 1011)

A very similar view has been expressed by a number of religious organizations. For example, as early as 1997, the Pontifical Academy for Life of the Roman Catholic Church issued an extended document explaining the reasons that human cloning was theologically and philosophically unacceptable. That document argued that

> "In the cloning process the basic relationships of the human person are perverted: filiation, consanguinity, kinship, parenthood. . . . As in every artificial activity, what occurs in nature is "mimicked" and "imitated", but only at the price of ignoring how man surpasses his biological component, which moreover is reduced to those forms of reproduction that have characterized only the biologically simplest and least evolved organisms." (Vial Correa 1997)

- "Human cloning would divert resources from other more important social and medical needs." It seems obvious that the world is faced with a multitude of medical and health problems that place a severe demand on the financial resources needed to deal with these problems. Human cloning of all kinds still tends to be very expensive, and to the extent that financial resources are directed to cloning, to the same extent will those resources be less available to dealing with that vast array of issues (McDougal 2008).

- "Human cloning might be used by commercial interests for financial gain." In capitalist societies, the opportunity to make a profit on one's efforts is a paramount concern. The conflict between profit making and individual rights and freedoms is an ongoing issue in such societies, so it is hardly surprising that concerns about profit making are embedded in the debate over human reproductive cloning. Obviously there is a significant potential for entrepreneurs to make very large profits in offering a variety of reproductive cloning opportunities to the general public. Such a scenario is even more likely when many governments ban the use of taxpayer funds to pay for cloning research and private companies have the opportunity to take over the cost (and potential profits) of such research. An early indication of such a potential dates to 2002, when the organization Patent Watch found that the U.S. Patent and Trademark Office had issued a patent to the University of Missouri for the production of cloned mammals, a procedure that, some say, could be used for the reproductive cloning of humans (Pollack 2002).

- "Human cloning might be used by governments or other groups for immoral and exploitative purposes." Literature is rife with scientific fiction tales of the possibilities created if and when immoral governments choose to use technology to produce specifically designed types of human beings to serve designated roles in society. Perhaps the classic example of this story line is Aldous Huxley's *Brave New World* in which each child is born genetically predisposed to take a designated role in the culture . . . and to be completely happy in doing so. Such stories usually (but not always) include an ending that entails warnings about the moral, ethical, psychological, and social horrors of such practices. But the stories continue to appear (Huxley 1932; Kass 2001).

- "Human cloning used on a very widespread basis would have a disastrous effect on the human gene pool by reducing genetic diversity and our capacity to adapt to new conditions." In his review, Brock points out that this concern has been raised by some writers, but that it is almost certainly one that should not receive much attention because human reproductive cloning "would not be used on a wide enough scale, substantially replacing sexual reproduction, to have the feared effect on the gene pool" (Brock 1997, E-20). (A number of print and electronic resources summarize arguments for and against human reproductive cloning. See, for example, Devolder 2014; Manninen 2014; Reproductive Cloning Arguments Pro and Con 2006; Wynn 2014.)

Therapeutic Cloning

As noted in Chapter 1, therapeutic cloning is a procedure that is carried out for the purpose of harvesting embryonic stem cells that can be used for the growth of differentiated cells, tissues, organs, or other biological products that can be used to treat a variety of human diseases and disorders. The procedure by which therapeutic cloning is carried out is the same as that for reproductive cloning, but the ultimate goal of the procedure is entirely different. No attempt is made to produce a complete organism, such as a new lamb, mouse, or human child. A major benefit of therapeutic cloning is that a patient's body is not likely to reject cells, tissue, or other products produced by this form of cloning because the embryonic stem cells from which those materials developed came from the patient's own body (e.g., tissue from the patient's skin). A second benefit is that the embryonic stem cells harvested for the procedure are pluripotent, capable of developing into almost any cell in the human body. Thus, they can be used to treat diseases and disorders of the blood, the heart, the nervous system, the liver, the respiratory system, or almost any other type of tissue or organ.

The discovery of human embryonic stem cells and the development of the technologies needed to work with them have opened up a whole new field of medicine, often known as *regenerative medicine*. An article published by the Mayo Clinic's Center for Regenerative Medicine claims that the new discipline "is a game-changing area of medicine with the potential to fully heal damaged tissues and organs, offering solutions and hope for people who have conditions that today are beyond repair" (About Regenerative Medicine 2014). Some of the many possible applications of therapeutic cloning and regenerative medicine include the following:

- The differentiation of embryonic stem cells into endocrine cells that would produce most of the pancreatic hormones (e.g., D'Amour et al. 2006)
- The synthesis of cells, blood vessels, and other structures that can be used in the circulatory system to treat atherosclerosis, infarcts, and other cardiac problems (e.g., Chen, Priest, and Gold 2008)
- The treatment of neurodegenerative diseases such as Alzheimer's disease, Parkinson's disease, Huntington's diseases, and amyotrophic lateral sclerosis (e.g., Tabar et al. 2008)
- The synthesis of tissue to replace materials lost through burns, injuries, disease, or other events (e.g., Atala and Koh 2004)
- The production of liver cells that may be able to correct for insulin inadequacy and, therefore, offer a cure for diabetes (e.g., Ben-Yeuhdah et al. 2004)
- The treatment of skin cancer and, presumably, other types of cancer (e.g., Hunder et al. 2008)
- The possible reversal of the aging process of cells, suggesting that therapeutic cloning might have the potential for extending a person's life span (e.g., Lanza et al. 2000)

(Lack of space prevents an extended discussion of the many possible applications and uses of therapeutic cloning. For more information on the topic, see Baharvand and Aghdami 2012; Jensen 2014; Rodriguez, Ross, and Cibelli 2012; Warburton 2015.)

The debate over human cloning over the past two decades has been among the most contentious in American society and many other nations around the world. Since the possibilities of human reproductive cloning first became apparent in the late 1990s, nearly all public opinion polls have shown a very large majority of opponents to the use of cloning technology to produce new human beings. Although polling organizations and the methods they use tend to differ significantly from survey to survey, the percentage of respondents opposing human reproductive cloning has consistently fallen in the range of 80 to 90 percent. That result does not appear to differ very much from nation to nation, between sexes, among various age groups, and within nearly all religious denominations (CGS Summary of Public Opinion Polls 2014).

By contrast, public opinion about therapeutic cloning has been very different during the same time period, with considerably more variation among subgroups. During the early part of this period, in 2002, support for the use of stem cells for therapeutic cloning research was at a low of about 35 percent. Over time, that percentage continued to rise until it reached a high in the mid-60 percent range. Support for therapeutic cloning was even higher—reaching the mid-70 percent level—when the question asked to respondents specifically excluded the use of human embryos for such research (Science and Engineering Indicators 2012, 7-40–7-42; for attitudes in Europe, see Biotechnology Report 2010, 97–114).

A number of arguments have been advanced in opposition to therapeutic cloning, such as the cost and pain and suffering that may be associated with the production and/or harvesting of human embryos, the low rate of success currently experienced in such procedures, and the inadequate number of human embryos available for use in such procedures. But by far the most serious problem in the advance of therapeutic cloning has been an ethical and theological question about the viability of an embryo used in such research. Many theologians and ethicists argue that human life begins at conception, at the very moment that a sperm enters and fertilizes an egg. For individuals who accept this position, the embryo that is to be used in therapeutic cloning is already a human being or, at the very least, a "potential" human being. Therefore, using the embryo to harvest stem cells for use in therapeutic cloning is actually an act of homicide and cannot be allowed under any circumstances.

Religious opposition to therapeutic cloning using embryonic stem cells runs through essentially all denominations. For example, Pope Benedict XVI reminded participants of a symposium on stem cell research in September 2006 that

> If there has been resistance [to stem cell research]—and if there still is—it was and is to those forms of research that provide for the planned suppression of human beings who already exist even if they have not yet been born. Research, in such cases, irrespective of efficacious therapeutic results is not truly at the service of humanity. (Address of His Holiness Benedict XVI to the Participants in the Symposium on the Theme: "Stem Cells: What Future for Therapy?" Organized by the Pontifical Academy for Life 2006)

The views of various religious denominations and theological leaders within those denominations often differ with regard to therapeutic cloning. One survey conducted by the Pew Research Religion and Public Life Project found that relatively few religious groups hold outright opposition to stem cell research under all circumstances, as is the case with the Roman Catholic Church. Most denominations have not expressed any formal view or take the position that stem cell research is permissible (and, in a few cases, even required) under certain conditions. The most common of those conditions is that the stem cells used in such research were not created specifically for the purpose of research, but were produced for some other purpose (e.g., artificial conception by a married couple) and would have been destroyed anyway (Religious Groups' Official Positions on Stem Cell Research 2008; among the many good references on the ethics of stem cell research, see Devolder 2015; De Wert and Mummery 2003; Embryonic Stem Cell Research 2011; Siegel 2013).

Probably the most common objection to therapeutic cloning, then, is concern over the production of human embryos that will survive no more than a few days, at the end of which time they will be harvested for their embryonic stem cells, and then destroyed. The ethical debate over this type of research stymied progress on therapeutic cloning for more than a decade, to the point that researchers began to look for other ways of carrying out the process that does not involve the use of human embryos. Thus far, two possible alternative technologies have been developed, the cloning of embryonic stem cells from adult somatic cells and a process known as induced pluripotent stem cells, or iPSCs.

The logic behind using adult somatic cells for therapeutic cloning is that this line of research is very unlikely to generate the strong opposition to therapeutic cloning generated by the use of human embryonic stem cells. This type of research begins by removing differentiated somatic cells from some part of an adult human body, often the skin, and tricking those cells into reverting to the earliest stage of their existence: the embryonic stem cells from which they originally grew. No ethical issue is likely to arise from such research because no human embryos are destroyed in the process. True, a person loses some skin cells, but that problem is not likely to generate a firestorm of protest.

Describing this process is simple, but making it happen is another matter. Until very recently, scientists had no idea as to how they might convert the character of an adult somatic cell and make it start behaving as a stem cell. That problem was finally solved in 2013 when researchers under the direction of Shoukhrat Mitalipov at Oregon Health Sciences University announced that they had successfully cloned human embryonic stem cells produced by somatic cell nuclear transfer (SCNT) procedures identical to those used in the cloning of Dolly and countless other animals. The difference in Mitalipov's research was that he had found a way to convince the fertilized egg produced by SCNT to start behaving as an embryo. In fact, it began to divide through a series of stages typical of human embryogenesis, reaching the blastomere stage at which stem cells could be harvested. And nothing was harmed except for some volunteer skin cells (Cyranoski 2013; Tachibana et al. 2013).

Induced pluripotency has been known for a somewhat longer period of time, having been first discovered in 2006 by Japanese researchers Shinya Yamanaka and his colleagues at Kyoto University and the University of California, San Francisco. (Yamanaka received a share of the 2012 Nobel Prize in Medicine or Physiology for his discovery.) Yamanaka's solution to the problem of therapeutic cloning was completely different from that of Mitalipov's approach in that it involved no transgenic eggs; no cloning; and, most important, no destruction of embryos. The process again sounds simple in concept, although it was by no means simple to accomplish. Somatic cells are removed from some part of an adult's body, with the skin again being a common donor site. Certain regulatory genes (now called *Yamanaka factors*) are then transfected into these cells. The genes are those known to have the ability to reprogram the cell and make it think that it is in an embryonic state; that is, they convert the differentiated adult cell to an undifferentiated stem cell, which then begins to replicate as would any embryonic stem cell. The stem cells then begin to replicate and are available as the reservoir for a variety of lines of stem cell research (Takahashi and Yamanaka 2006; With Few Factors, Adult Cells Take on Character of Embryonic Stem Cells 2006).

Cloning Laws and Legislation

Legislation dealing with cloning research in the United States dates to the mid-1970s and arose largely in response to the Supreme Court's decision in the case of *Roe v. Wade*. In that case, the Court's ruling that a woman could choose to have an abortion up until the end of the second trimester implied that the fetus had no rights to survival. In response to that ruling many Right to Life groups initiated efforts to prevent researchers from using fetuses from being treated as "objects" rather than living humans in their studies. They began to lobby Congress to tighten restrictions on the use of fetal tissue for research

(Wertz 2002; for a detailed discussion of the early history of cloning legislation in the United States, see Fletcher 2000).

Partly in response to this pressure, Congress passed the National Research Act of 1974, which, among other things, established a National Commission for the Protection of Human Subjects in Biomedical and Behavioral Research to study issues related to fetal research and placed a moratorium on such research until the commission had reported back to Congress. A year later, the commission recommended that all fetuses be treated equally in research and that, in particular, a standard of "minimal risk" be established, meaning that the fetus be exposed to no risk "greater in and of themselves than those ordinarily encountered in daily life or during the performance of routine physical or psychological examinations or tests" (Vawter 1993).

The action taken by Congress in 1974 essentially placed a ban on the use of federal funds for fetal research, a ban that has remained in place ever since. During that period of time, both the Congress and various U.S. presidents have taken additional actions with regard to the funding of fetal research, but federal policy remains essentially the same: No federal funds can be spent on fetal research, although there remains no federal law preventing the funding of such research by private sources. Some specific events that have marked the debate over federal funding of stem cell research include the following:

1985: Congress passed the Health Research Extension Act of 1985 (Public Law 99-158), which said that no harm of any kind (the so-called golden rule standard) could be caused to a fetus during research, essentially extending the ban on fetal research for cloning purposes (Fletcher 2000, E-10; Public Law 99-158, 1985).

1993: President Bill Clinton, in one of his first acts in office, directed the Secretary of Health and Human Services to lift the ban on the federal funding of human stem cell research for therapeutic purposes. Shortly thereafter, the

Congress passed the National Institutes of Health Revitalization Act of 1993, legislatively confirming Clinton's executive action (Flannery and Javitt 2000, D-4–D-5; Public Law 103-43, 1993).

1994: The Human Embryo Research Panel appointed a year earlier by Clinton recommended that the federal government continue funding research on therapeutic cloning. Clinton, however, overruled the panel's recommendation (Kolata 1997).

1995: Dissatisfied with President Clinton's actions on the federal funding of human embryonic stem cell research, Congress adopted the so-called Dickey (or Dickey-Wicker) Amendment to the appropriation bills for the Departments of Education and Health and Human Services. The amendment banned the use of federal funds for creation of or research on human embryos for any purpose whatsoever. The ban remained in effect until it was overturned by a decision of the U.S. Court of Appeals for the District of Columbia in 2011 (Kaiser 2011; Kearl 2013).

2001: President George W. Bush, attempting to clarify the status of stem cell research in the United States, announced that such research would be banned except for studies on 64 stem cell lines that were already in existence and available for research. Bush's action was challenged on a number of occasions by the Democratic-controlled Congress, always without success. The Stem Cell Research Enhancement Act was first introduced in 2005 and then again in 2007 and 2009. Various versions of the bill were either not passed by the Congress or eventually vetoed by the president. (For a review of the type of stem cell research legislation offered during Bush's presidency, see Human Embryonic Stem Cells Research 2014.)

2009: In one of his first major acts in office, President Barack Obama issued Executive Order 13505, which reversed President George W. Bush's eight-year policy on human embryonic research. The order instructed the National Institutes of Health to prepare guidelines for the funding of such research. During President Obama's term, the federal government has continued to move forward with the funding of human embryonic stem cell research. As this history has shown, however, past history is no indicator as to the direction that federal funding for therapeutic cloning may take in the future. (For information on current federal policy and actions on the funding of human embryonic stem cell research see Stem Cell Information 2014.)

Since 2009, members of the Congress have also attempted to pass legislation that would provide a firmer basis for stem cell research than that offered by executive orders, such as those issued by President Obama. For example, in 2011, Representative Diana DeGette (D-CO) introduced H.R. 2376, the Stem Cell Research Advancement Act, to achieve this goal. The bill had 30 cosponsors, but never made it out of committee. Representative DeGette reintroduced the bill again in the 113th Congress in 2013, this time with two cosponsors. Again, the bill failed to go beyond the committee level (Reed 2010).

State Laws

Individual states have been considerably more aggressive than the federal government about adopting cloning laws. To date, 18 states have adopted legislation dealing with either reproductive or therapeutic cloning, or both. Laws in Arizona, Arkansas, Indiana, Michigan, North Dakota, Oklahoma, South Dakota, and Virginia ban both type of cloning, whereas laws in California, Connecticut, Illinois, Iowa, Maryland, Massachusetts, Missouri, Montana, New Jersey, and Rhode Island ban only reproductive cloning, but not therapeutic cloning. These laws differ somewhat from state to state, with some legislation banning all forms of cloning outright, others prohibiting only certain types of cloning, and still others restricting only the use of state funds for cloning research. Two other states have somewhat different types of laws that prohibit healthcare workers from participating in any type of cloning research that violates their own conscience. (A law dealing with experimentation on a "living human conceptus" may or may not apply specifically to cloning research; Summary of State Laws on Human Cloning 2011; for more detail on individual state laws, see Human Cloning Laws 2011; States 2014.)

Five states also have laws that specifically permit the use of state funds for cloning research. These laws were all passed in 2004 and 2005 when enthusiasm for the medical and health potential for therapeutic cloning was at a peak. The first of these laws was adopted by the New Jersey state legislature in June 2004. That law established the New Jersey Stem Cell Institute, created to promote and carry out therapeutic cloning and regenerative medicine projects. Five months later, California voters voted in favor of Proposition 71, which provided $3 billion for the funding of cloning research in the state over a period of 10 years. Similar legislation was later adopted in Connecticut, Missouri, New York, and Rhode Island. These states now collectively provide more funding for stem cell research than does the federal government (Levine 2010; for further information on types of state stem cell funding programs, also see State Initiatives for Stem Cell Research 2009).

International Policies

In contrast to the United States, most other countries of the world have adopted laws specifically prohibiting reproductive cloning and, in many cases, therapeutic cloning also. According to some of the most recent data available, laws against reproductive cloning have been adopted in at least 40 countries and against therapeutic cloning, in at least 18 nations. The former category includes a wide range of nations, including India, Vietnam, Turkey, Tunisia, Panama, and Iceland, as well as all of North America and most of Europe. Bans on therapeutic cloning tend to occur primarily in North American and the European Union (Wheat and Matthews 2007).

According to some observers, the regulation of human cloning is prohibited implicitly in the Council of Europe Convention on Human Rights and Biomedicine, which provides basic protection for all humans, with the definition of *humans* being left essentially to each member state. All of the EU states have now signed this convention (Human Cloning Regulation in Europe 2004). A more specific piece of legislation, the Protocol on the Prohibition of Cloning Human Beings, entered into force in March 2001 when it had been ratified by five nations. The protocol specifically banned "any intervention seeking to create a human being genetically identical to another human being, whether living or dead," which observers have interpreted as banning reproductive cloning. That treaty has now been signed and ratified by 15 European nations, few of them major members of the union, such as France, Germany, Italy, or the United Kingdom. It has also been largely ignored by other nations eligible to sign the protocol, such as Australia, Canada, Japan, the Holy See, and the United States (Additional Protocol 2005).

Conclusion

The technology used in cloning animals has been around for little more than two decades. Yet the debate over the proposed use—or avoidance of use—of this technology has already reached a fever pitch in some specific areas, such as the reproductive cloning of humans and the production of food products from cloned animals. One can scarcely imagine the advances in technology that may occur in the next two decades or more or in the ways in which these advances also might be put to use. The technical ability to clone a human being, for example, can become more and more likely only with the consequent impact that change will have on the debate over reproductive cloning. It seems inevitable, then, that as exciting as the technological progress in cloning is likely to be, so must the trepidation with which humans approach the social, political, and ethical issues created by this progress.

References

"1 Year Later: Cat, Clone Differ." 2003. USA Today. http://usatoday30.usatoday.com/news/science/2003-01-21-cloned-cats_x.htm. Accessed on December 17, 2014.

"150K to Clone a Pet?" 2014. Let's Talk Bible. http://bibleforums.org/showthread.php/156534-150k-To-clone-Pet?s=5f03047679b1182972be61ac1a3844c1. Accessed on December 17, 2014.

"About Regenerative Medicine." 2014. Mayo Clinic. http://www.mayo.edu/research/centers-programs/center-regenerative-medicine/patient-care/about-regenerative-medicine. Accessed on December 22, 2014.

"Address of His Holiness Benedict XVI to the Participants in the Symposium on the Theme: 'Stem Cells: What Future for Therapy?' Organized by the Pontifical Academy for Life." 2006. http://w2.vatican.va/content/benedict-xvi/en/speeches/2006/september/documents/hf ben-xvi spe 20060916 pav.html. Accessed on May 13, 2015.

"All about Animal Cloning." 2010. Animal Biotechnology Update. https://www.bio.org/sites/default/files/Cloning_onepager.pdf. Accessed on December 16, 2014.

"American Association for the Advancement of Science Statement on Human Cloning." 2013. http://www.aaas.org/page/american-association-advancement-science-statement-human-cloning. Accessed on December 21, 2014.

"Animal Cloning." 2014. Food and Drug Administration. http://www.fda.gov/AnimalVeterinary/SafetyHealth/AnimalCloning/. Accessed on December 17, 2014.

"Animal Cloning: A Risk Assessment." 2008. Center for Veterinary Medicine. U.S. Food and Drug Administration. http://www.fda.gov/downloads/AnimalVeterinary/SafetyHealth/AnimalCloning/UCM124756.pdf. Accessed on December 17, 2014.

Appel, Jacob M. 2009. "Should We Really Fear Human Reproductive Cloning?" Huffington Post. http://www.huffingtonpost.com/jacob-m-appel/ should-we-really-fear-rep_b_183308.html. Accessed on December 20, 2014.

Appleton, Caroline. 2013. "The First Successful Cloning of a Gaur (2000), by Advanced Cell Technology." The Embryo Project Encyclopedia. http://embryo.asu.edu/pages/ first-successful-cloning-gaur-2000-advanced-cell-technol ogy. Accessed on December 18, 2014.

"Applications of Cell and Tissue Culture." 2014. Biotechnology 4U. http://www.biotechnology4u.com/plant_biotech nology_applications_cell_tissue_culture.html. Accessed on December 12, 2014.

"Asexual Reproduction in Plants: Advantages, Disadvantages & Types." 2014. Education Portal. http://educa-tion-portal.com/academy/lesson/asexual-reproduc tion-in-plants-advantages-disadvantages-types.html. Accessed on December 12, 2014.

Atala, Anthony, and Chester J. Koh. 2004. "Tissue Engineering Applications of Therapeutic Cloning." *Annual Review of Biomedical Engineering*. 6(1): 27–40.

"Back from the Dead: Couple Who Paid $155,000 to Clone Their Dog after His Sudden Death." 2012. Mail Online. http://www.dailymail.co.uk/news/article-2141574/ First-couple-clone-pet-dog-paid-155-000-job-say-ne w-pup-mannerisms-dead-Labrador.html. Accessed on December 17, 2014.

Baharvand, Hossein, and Nasser Aghdami, eds. 2012. *Regenerative Medicine and Cell Therapy*. Amsterdam: IOS Press.

Barrett, Rick. 2010. "Medicine, Not Food, May Have More to Gain from Cloning." Milwaukee-Wisconsin Sentinel Journal. http://www.jsonline.com/news/health/100703769 .html. Accessed on December 16, 2014.

Ben-Yeuhdah, Ahmi, et al. 2004. "Can Diabetes Be Cured by Therapeutic Cloning?" *Pediatric Diabetes.* 5: 79–87.

"Biotechnology Report." 2010. Eurobarometer. European Commission. http://ec.europa.eu/public_opinion/archives/ebs/ebs_341_en.pdf. Accessed on December 23, 2014.

"Bos gaurus." 2014. The IUCN Red List of Endangered Species. http://www.iucnredlist.org/details/2891/0. Accessed on December 18, 2014.

Brand, Stewart. 2014. "Frequently Asked Questions." Revive & Restore. http://longnow.org/revive/faq/. Accessed on December 19, 2014.

Brand, Stewart. 2013. "Opinion: The Case for Reviving Extinct Species." National Geographic. http://news.nationalgeographic.com/news/2013/03/130311-deextinction-reviving-extinct-species-opinion-animals-science/. Accessed on December 19, 2014.

Breitowitz, Yitzchok. 2002. "What's So Bad about Human Cloning?" *Kennedy Institute of Ethics Journal.* 12(4): 325–341.

Brock, Dan W. 1997. "Cloning Human Beings: An Assessment of the Ethical Issues Pro and Con." https://repository.library.georgetown.edu/bitstream/handle/10822/559359/nbac_cloning2.pdf?sequence=1. Accessed on December 19, 2014, E-1–E-24.

Burr, Benjamin, and Frances Burr. 2002. "How Do Seedless Fruits Arise, and How Are They Propagated?" *Scientific American.* 286(1): 98.

Carroll, Sarah. 2014. "The Pros and Cons of Plant Cloning." eHow. http://www.ehow.co.uk/info_7937048_pros-cons-plant-cloning.html. Accessed on December 14, 2014.

Cattle, Daniel, Thomas McCreanor, and Lawrence Mago. 2008. "Plant Cloning." http://www.slideshare.net/roorensu/plant-cloning-presentation-413164. Accessed on December 12, 2014.

"Cattle Cloning." 2014. Utah State University. http://www
.biosystems.usu.edu/htm/research/cloning/cattle-
cloning#dairy. Accessed on December 16, 2014.

"CGS Summary of Public Opinion Polls." 2014. Center
for Genetics and Society. http://www.geneticsand
society.org/article.php?id=401#research. Accessed on
December 23, 2014.

Chen, Yinhong, Catherine A. Priest, and Joseph D. Gold.
2008. "Myocardial Regenerative Potential by Stem Cell
Transplant." Chap. 27 in Ronald Ross Watson and Douglas
F. Larson, eds. *Immune Dysfunction and Immunotherapy in
Heart Disease.* Oxford, UK: Blackwell Futura.

Cheshire, William P., et al. 2003. "Stem Cell Research: Why
Medicine Should Reject Human Cloning." *Mayo Clinic
Proceedings.* 78: 1010–1018. http://www.mayoclinicpro
ceedings.org/article/S0025-6196%2811%2963148-0/pdf.
Accessed on December 21, 2014.

Chong, S. T., and T. B. Chai. 1990. "Recent Developments
in Vegetative Propagation of Some Tropical Fruit Trees."
The Archives of the Rare Fruit Council of Australia. http://
rfcarchives.org.au/Next/CaringForTrees/Grafting5-90.htm.
Accessed on December 12, 2014.

"Citizen Petition." 2014. Consumers Union. http://www.
fsis.usda.gov/wps/wcm/connect/6122594c-93db-46db-b
eb6-dc250bc43b6d/Petition-Consumers-Union-062614.
pdf?MOD=AJPERES. Accessed on December 17, 2014.

"Clonaid Claims Were a Hoax." 2003. BioNews. http://
www.bionews.org.uk/page_11748.asp. Accessed on
December 12, 2014.

"Cloning Human Beings." 1997. National Bioethics Advisory
Committee. Rockville, Maryland, June 1997. https://bio
ethicsarchive.georgetown.edu/nbac/pubs/cloning1/cloning.
pdf. Accessed on December 19, 2014.

"Cloning Lawsuit." 2013. American Quarter Horse Association. http://www.aqha.com/News/News-Articles/2013/

July/07302013-Lawsuit-Verdict.aspx. Accessed on December 16, 2014.

"Cloning Research Continues at USU." 2014. Cache Valley Daily.com. http://www.cachevalleydaily.com/news/local/ article_dd418581-17be-5337-a0c7-a33552eb538e.html. Accessed on December 16, 2014.

"Cloning Your Dog." 2014. My Friend Again. http://my-friendagain.com/Dog%20Cloning%20Cost/Dog%20 Cloning%20cost.htm. Accessed on December 17, 2014.

"Compilation and Analysis of Public Opinion Polls on Animal Cloning." 2008. The Center for Food Safety. http:// www.centerforfoodsafety.org/files/polling_animal_clone_ summary_2-20-08.pdf. Accessed on December 30, 2014.

Conger, Cristen. 2014. "Could We Clone Our Organs to Be Used in a Transplant?" How Stuff Works. http://science. howstuffworks.com/life/genetic/cloned-organ-transplant. htm. Accessed on December 20, 2014.

"Consequences of Low Genetic Diversity." 2014. The Teacher-Friendly Guide to the Evolution of Maize. http:// maize.teacherfriendlyguide.org/index.php/genetic-diversity-and-evolution/consequences-of-low-diversity. Accessed on December 14, 2014.

Cyranoski, David. 2013. "Human Stem Cells Created by Cloning." *Nature*. 497(7449): 295–296.

Cyranoski, David. 2014. "Cloning Comeback." *Nature*. 505(7484): 468–471. http://www.nature.com/news/cloning-comeback-1.14504. Accessed on December 17, 2014.

D'Amour, K.A., et al. 2006. "Production of Pancreatic Hormone-Expressing Endocrine Cells from Human Embryonic Stem Cells." *Nature Biotechnology*. 24(11): 1392–1401.

Dean, Josh. 2014. "For $100,000, You Can Clone a Dog." Bloomberg Businessweek. http://www.businessweek.com/ articles/2014-10-22/koreas-sooam-biotech-is-the-worlds-first-animal-cloning-factory. Accessed on December 17, 2014.

"Deaths of Transgenic Calves at AgResearch's Ruakura Facility." 2010. Report to Minister of Research, Science, and Technology. http://www.gefree.org.nz/assets/pdf/Gluckmanreportoncattledeaths.pdf. Accessed on December 16, 2014.

Delius, Charlie, and Patrick Keegan. 2014. "Layering and Grafting for Native Plant Propagation." http://depts.washington.edu/propplnt/Chapters/Layering%20and%20grafting.pdf. Accessed on December 12, 2014.

Devolder, Katrien. 2014. "Cloning." Stanford Encyclopedia of Philosophy. http://plato.stanford.edu/entries/cloning/. Accessed on December 16, 2014.

Devolder, Katrien. 2015. *The Ethics of Embryonic Stem Cell Research*. New York: Oxford University Press.

De Wert, Guido, and Christine Mummery. 2003. "Human Embryonic Stem Cells: Research, Ethics and Policy." *Human Reproduction*. 18(4): 672–682.

Dixon, Patrick. 2014. "What Is Human Cloning? How to Clone. But Ethical?" http://www.globalchange.com/clonech.htm. Accessed on December 20, 2014.

Doig, Will. 2003. "Born Again." Metro Weekly. http://www.metroweekly.com/2003/01/born-again/. Accessed on December 20, 2014.

Edge, Raymond S., and John Randall Groves. 2006. *Ethics of Health Care: A Guide for Clinical Practice*, 3rd ed. New York: Delmar.

"Embryonic Stem Cell Research: An Ethical Dilemma." 2011. EuroStemCell. http://www.eurostemcell.org/factsheet/embryronic-stem-cell-research-ethical-dilemma. Accessed on December 23, 2014.

"EP Says No to Food from Cloned Animals." 2014. BEUC (European Consumers Organisation). http://www.beuc.org/press-media/news-events/ep-says-no-food-cloned-animals. Accessed on December 30, 2014.

"Europeans' Attitudes toward Animal Cloning." 2008. Flash Eurobarometer. http://ec.europa.eu/public_opinion/flash/fl_238_en.pdf. Accessed on December 30, 2014.

Evans, M. J., and M. H. Kaufman. 1981. "Establishment in Culture of Pluripotential Cells from Mouse Embryos." *Nature.* 292(5819): 154–156.

Falk, Donald A., Eric Knapp, and Edgar O. Guerrant. 2001. "Why Is Genetic Diversity Important?" An Introduction to Restoration Genetics. http://www.nps.gov/plants/restore/pubs/restgene/1.htm. Accessed on December 13, 2014.

Feinberg, Joel, 1980. "A Child's Right to an Open Future." In William Aiken and Hugh LaFollette, eds. *Whose Child? Parental Rights, Parental Authority and State Power.* Totowa, NJ: Rowman & Littlefield, 124–153.

Finley, Bill. 2013. "The Sport of Kings, and Clones?" ESPN Horse Racing. http://espn.go.com/horse-racing/story/_/id/9582361/the-sport-kings-clones. Accessed on December 16, 2014.

Flannery, Ellen J., and Gail H. Javitt. 2000. "Analysis of Federal Laws Pertaining to Funding of Human Pluripotent Stem Cell Research." *Ethical Issues in Human Stem Cell Research. Vol. II.* Rockville, MD: National Bioethics Advisory Commission, D-1–D-13. https://bioethicsarchive.georgetown.edu/nbac/stemcell2.pdf. Accessed on December 27, 2014.

Fletcher, John C. 2000. "Deliberating Incrementally on Human Pluripotent Stem Cell Research." *Ethical Issues in Human Stem Cell Research.* Vol. II. Rockville, MD: National Bioethics Advisory Commission, E-1–E-50. https://bioethicsarchive.georgetown.edu/nbac/stemcell2.pdf. Accessed on December 27, 2014.

Folch, J., et al. 2009. "First Birth of an Animal from an Extinct Subspecies *(Capra Pyrenaica Pyrenaica)* by Cloning." *Theriogenology.* 71(6): 1026–1034.

Friese, Carrie. 2013. *Cloning Wild Life: Zoos, Captivity, and the Future of Endangered Animals*. New York: New York University Press.

Furcht, Leo, and William R. Hoffman. 2011. *The Stem Cell Dilemma: The Scientific Breakthroughs, Ethical Concerns, Political Tensions, and Hope Surrounding Stem Cell Research*, 2nd ed. New York: Arcade Publishers.

"Genetic Savings & Clone, Inc." 2014. Internet Archive Wayback Machine. http://web.archive.org/web/20060429204721/http://savingsandclone.com/index.html. Accessed on December 17, 2014.

Gibson, Eloise. 2010. "Mutant Cows Die in GM Trial." New Zealand Herald. http://www.nzherald.co.nz/nz/news/article.cfm?c_id=1&objectid=10642031. Accessed on December 16, 2014.

Gillam, Lynn. 2008. "Cloning to Avoid Genetic Disease." In Loane Skene and Janna Thompson, eds. *The Sorting Society: The Ethics of Genetic Screening and Therapy*. New York: Cambridge University Press, 37–50.

Goforth, David. 2014. "Advantages and Disadvantages to Asexual Propagation of Plants." http://www4.ncsu.edu/~djgofort/Miscellaneous4.htm. Accessed on November 14, 2014.

Goldenberg, Suzanne, and Alok Jha. 2004. "The World's First Cloned Pet (Cost $50,000)." The Guardian. http://www.theguardian.com/world/2004/dec/24/sciencenews.genetics. Accessed on December 17, 2014.

Gómez, M. C., et al. 2004. "Birth of African Wildcat Cloned Kittens Born from Domestic Cats." *Cloning Stem Cells*. 6(3): 247–258.

Haldane, J. B. S. 1963. "Biological Possibilities for the Human Species in the Next Ten Thousand Years." In G. E. W. Wolstenholme, ed. *Man and His Future*. Boston: Little, Brown.

Hare, Doug. 2003. "What of Animal Cloning?" *Canadian Veterinary Journal.* 44(4): 271–272.

Herren, Ray V. 2013. *Introduction to Biotechnology: An Agricultural Revolution*, 2nd ed. Clifton Park, NY: Delmar, Cengage Learning.

Hershey, David. 2002. "What Are Some Good Things and Some Bad Things about Plant Cloning?" http://www.madsci. org/posts/archives/2002-02/1014254397.Bt.r.html. Accessed on November 14, 2014.

Higgins, Charlie. 2014. "The Advantages of Plant Cloning." http://www.ehow.com/about_5480031_advantages-plant-cloning.html. Accessed on December 12, 2014.

"History." 2014. Sooam Biotech Research Foundation. http://en.sooam.com/about/sub03.html. Accessed on December 17, 2014.

Holguin, Jaime. 2002. "Milking Cows for All They're Worth." CBS News. http://www.cbsnews.com/news/milking-cows-for-all-theyre-worth/. Accessed on December 16, 2014.

"Human Cloning Laws: 50 State Survey." 2011. Bioethics Defense Fund. http://bdfund.org/wordpress/wp-content/uploads/2012/07/CLONINGChart-BDF2011.docx.pdf. Accessed on December 29, 2014.

"Human Cloning Regulation in Europe." 2004. American Center for Law and Justice. http://184.106.84.133/pro-life/human-cloning-regulation-in-europe. Accessed on December 29, 2014.

"Human Embryonic Stem Cells Research." 2014. Office of Legislative Policy and Analysis. http://olpa.od.nih.gov/legislation/109/pendinglegislation/embryonic.asp. Accessed on December 28, 2014.

Hunder, Naomi, N. et al. 2008. "Treatment of Metastatic Melanoma with Autologous CD4+ T Cells against NY-ESO-1." *New England Journal of Medicine.* 358(25): 2698–2703.

Huxley, Aldous. 1932. *Brave New World*. London: Chatto & Windus.

Hwang, I., et al. 2013. "Successful Cloning of Coyotes through Interspecies Somatic Cell Nuclear Transfer Using Domestic Dog Oocytes." *Reproduction, Fertility, and Development*. 25(8): 1142–1148.

Inbar, Michael. 2011. "Born to Save Sister's Life, She's Glad I Am in This Family." Today. http://www.today.com/ id/43265160/ns/today-today_news/t/born-save-sisters-life-shes-glad-i-am-family/#.VJYEy14AKA. Accessed on December 20, 2014.

Jabr, Ferris. 2013. "Will Cloning Ever Save Endangered Animals?" *Scientific American*. http://www.scientificamerican. com/article/cloning-endangered-animals/. Accessed on December 18, 2014.

Jensen, Eric Allen. 2014. *The Therapeutic Cloning Debate: Global Science and Journalism in the Public Sphere*. Burlington, VT: Ashgate.

Jiang, Caifu, et al. 2011. "Regenerant Arabidopsis Lineages Display a Distinct Genome-Wide Spectrum of Mutations Conferring Variant Phenotypes." *Current Biology*. 21(16): 1385–1390. http://www.ncbi.nlm.nih.gov/pmc/articles/ PMC3162137/. Accessed on December 14, 2014.

Kaiser, Jocelyn. 2011. "Stem Cell Ruling a Decisive Victory for NIH." ScienceInsider. http://news.sciencemag. org/2011/07/stem-court-ruling-decisive-victory-nih. Accessed on December 28, 2014.

Kass, Leon. 2001. "Preventing a Brave New World." http:// web.stanford.edu/~mvr2j/sfsu09/extra/Kass3.pdf. Accessed on December 22, 2014.

Kass, Leon, ed. 2002. *Human Cloning and Human Dignity: The Report of the President's Council on Bioethics*. President's Council on Bioethics. New York: Public Affairs.

Kearl, Megan. 2013. "Dickey-Wicker Amendment." The Embryo Project Encyclopedia. http://embryo.asu.

edu/pages/dickey-wicker-amendment. Accessed on December 28, 2014.

King, David. 1999. "Human Cloning—Without Our Consent." Aisling Magazine. http://www.aislingmagazine.com/aislingmagazine/articles/TAM25/HumanCloning.html. Accessed on December 16, 2014.

Kolata, Gina. 1997. "Little-Known Panel Challenged to Make Quick Cloning Study." New York Times. https://partners.nytimes.com/library/national/0318sci-clone-ethics.html. Accessed on December 28, 2014.

Kutner, Jennie. 2014. "UK Woman Wins Competition to Have Dachsund Cloned for Free." The Dodo. https://www.thedodo.com/uk-woman-wins-competition-to-h-503939183.html. Accessed on December 17, 2014.

Langwith, Jacqueline, ed. 2012. *Cloning*. Detroit: Greenhaven Press/Gale Cengage Learning.

Lanza, Robert P., Betsy L. Dresser, and Philip Damiani. 2000. "Cloning Noah's Ark." *Scientific American*. 283(5): 84–89.

Lanza, Robert P., et al. 2000. "Extension of Cell Life-span and Telomere Length in Animals Cloned from Senescent Somatic Cells." *Science*. 288(5466): 665–669.

Levine, Aaron. 2010. "States Now Fund a Majority of Human Embryonic Stem Cell Research." Georgia Tech. http://www.news.gatech.edu/2010/12/09/states-now-fund-majority-human-embryonic-stem-cell-research. Accessed on December 29, 2014.

"Life-Span." 2014. The Free Dictionary. http://encyclopedia2.thefreedictionary.com/Life-Span. Accessed on December 14, 2014.

Loi, P., et al. 2001. "Genetic Rescue of an Endangered Mammal by Cross-Species Nuclear Transfer Using Post-Mortem Somatic Cells." *Nature Biotechnology*. 19(10): 962–964.

Loxdale, Hugh D. 2009. "What's in a Clone: The Rapid Evolution of Aphid Asexual Lineages in Relation to Geography,

Host Plant Adaptation and Resistance to Pesticides." *In* Isa Schön, Koen Martens, and Peter Dijk, eds. *Lost Sex: The Evolutionary Biology of Parthenogenesis.* New York: Springer.

Mameli, M. 2007. "Reproductive Cloning, Genetic Engineering and the Autonomy of the Child: The Moral Agent and the Open Future." *Journal of Medical Ethics.* 33(2): 87–93. http://www.ncbi.nlm.nih.gov/pmc/articles/PMC2598241/. Accessed on December 21, 2014.

Manninen, Bertha Alvarez. 2014. "Cloning." Internet Encyclopedia of Philosophy. http://www.iep.utm.edu/cloning/. Accessed on December 14, 2014.

Marcondes, Alice. 2012. "Brazil Embarks on Cloning of Wild Animals." TerraAmérica. http://www.ipsnews.net/2012/11/brazil-embarks-on-cloning-of-wild-animals/. Accessed on December 18, 2014.

Mark, Jason. 2013. "Back from the Dead." Earth Island Journal. http://www.earthisland.org/journal/index.php/eij/article/back_from_the_dead/. Accessed on December 19, 2014.

Matulef, Kim. 2013. "Cloning." The Tech Museum. http://genetics.thetech.org/ask/ask147. Accessed on December 17, 2014.

Maugh, Thomas, II. 2011. "Ernest McCulloch Dies: Scientist Who First Identified Stem Cells Was 84." Washington Post. http://www.washingtonpost.com/wp-dyn/content/article/2011/02/06/AR2011020604018.html. Accessed on December 11, 2014.

McDougall, R. 2008. "A Resource-Based Version of the Argument That Cloning Is an Affront to Human Dignity." *Journal of Medical Ethics.* 34(4): 259–261.

"Monoculture and the Irish Potato Famine: Cases of Missing Genetic Variation." Understanding Evolution. http://evolution.berkeley.edu/evolibrary/article/agriculture_02. Accessed on December 14, 2014.

Miyao, Akio, et al. 2012. "Molecular Spectrum of Somaclonal Variation in Regenerated Rice Revealed by Whole-Genome Sequencing." *Plant & Cell Physiology*. 53(1): 256–264.

Morgenstern, Ken. 2011. "2nd Chance. This American Life." YouTube. https://www.youtube.com/watch?v=Obzfz4xEpgY. Accessed on December 17, 2014.

National Research Council, et al. 2010. *Final Report of the National Academies' Human Embryonic Stem Cell Research Advisory Committee and 2010 Amendments to the National Academies' Guidelines for Human Embryonic Stem Cell Research*. Washington, DC: National Academies Press.

Newton, David E. 2014. *GMO Food: A Reference Handbook*. Santa Barbara, CA: ABC-CLIO.

Nijman, I. J., et al. 2003. "Hybridization of Banteng (*Bos javanicus*) and Zebu (*Bos indicus*) Revealed by Mitochondrial DNA, Satellite DNA, AFLP and Microsatellites." *Heredity*. 90: 10–16.

Orentilcher, David. 2001. "Beyond Cloning: Expanding Reproductive Options for Same-Sex Couples." *Brooklyn Law Review*. 66(Part 3): 651–684.

Owen, Henry R. 2002. "The Cloning of Plants." The Keep. http://thekeep.eiu.edu/cgi/viewcontent.cgi?article=1144&context=bio_fac. Accessed on December 14, 2014.

Parsley, Larissa. 2004. "Transgenic Plants: A Budding Controversy Stems from Consumer Concerns." Journal of Young Investigators. http://www.jyi.org/issue/transgenic-plants-a-budding-controversy-stems-from-consumer-concerns. Accessed on December 14, 2014.

Pera, Martin F., Benjamin Reubinoff, and Alan Trounson. 2000. "Human Embryonic Stem Cells." *Journal of Cell Science*. 113(Pt 1): 5–10.

Perrin, Jacob, and Nancy King. 2014. "Ethical Issues in Stem Cell Research and Therapy." *Stem Cell Research & Therapy*.

5(4): article #85. http://stemcellres.com/content/5/4/85/ abstract. Accessed on December 21, 2014.

"Pet Cloning: Separating Facts from Fluff." 2005. American Anti-Vivisection Society. http://aavs.org/cms/assets/ uploads/2014/08/aavs_report_pet-cloning-facts-fluff. pdf?2e9aef. Accessed on December 17, 2014.

"Pharming for Farmaceuticals." 2014. Learn.Genetics. http://learn.genetics.utah.edu/content/science/pharming/. Accessed on December 16, 2014.

Pighin, Jamie. 2003. "Transgenic Crops: How Genetics Is Providing New Ways to Envision Agriculture." The Science Creative Quarterly. http://www.scq.ubc.ca/ transgenic-crops-how-genetics-is-providing-new-ways-to-envision-agriculture/. Accessed on December 14, 2014.

Pimm, Stuart. 2013. "Opinion: The Case against Species Revival." National Geographic. http://news.nationalgeo graphic.com/news/2013/03/130312—deextinction-conser vation-animals-science-extinction-biodiversity-habitat-envi ronment/. Accessed on December 19, 2014.

Pohlmeier, Bill, and Alison Van Eenennaam. 2008. "Biomedical Applications of Genetically Engineered and Cloned Animals." http://animalscience.ucdavis.edu/animalbiotech/ Outreach/Biomedical_applications_genetically_engineered_ animals.pdf. Accessed on December 16, 2014.

Poinar, Hendrik. 2013. "Bring back the Woolly Mammoth!" TED. https://www.ted.com/talks/hendrik_ poinar_bring_back_the_woolly_mammoth. Accessed on December 19, 2014.

Pollack, Andrew. 2002. "Debate on Human Cloning Turns to Patents." New York Times. http://www.nytimes. com/2002/05/17/science/17CLON.html. Accessed on December 22, 2014.

"Public Law 99-158." 1985. http://history.nih.gov/ research/downloads/PL99-158.pdf. Accessed on December 27, 2014.

"Public Law 103-43." 1993. http://history.nih.gov/ research/downloads/PL103-43.pdf. Accessed on December 28, 2014.

Quick, Susanne. 2005. "Copy Cats." Investigate. July: 88–89. http://issuu.com/iwishart/docs/investigate_july05. Accessed on December 17, 2014.

Quirós, Gabriela. 2007. "Reawakening Extinct Species." Quest. http://science.kqed.org/quest/video/reawakening-extinct-species/. Accessed on December 19, 2014.

Ramalho-Santos, Miguel, and Holger Willenbring. 2007. "On the Origin of the Term 'Stem Cell'." 1(1): 35–38.

Reed, David William. 2007. "Horticultural Workshops." Texas A&M University. http://generalhorticulture.tamu. edu/hort604/workshopmex07/propsoilwaterworkshop. htm. Accessed on December 12, 2014.

Reed, Don C. 2010. "Stem Cell Research Advancement Act?" Stem Cell Battles. http://stemcellbattles.wordpress. com/2010/10/27/stem-cell-research-advancement-act/. Accessed on December 29, 2014.

"Religious Groups' Official Positions on Stem Cell Research." 2008. Pew Research Religion and Public Life Project. http://www.pewforum.org/2008/07/17/religious-groups-official-positions-on-stem-cell-research/. Accessed on December 23, 2014.

"Reproductive Cloning." 2004. Religious Tolerance. http:// www.religioustolerance.org/clo_intra.htm#mor. Accessed on December 20, 2014.

"Reproductive Cloning: Ethical and Social Issues." 2004. Human Genetics Alert. http://www.hgalert.org/topics/ cloning/cloning.PDF. Accessed on December 20, 2014.

"Reproductive Cloning Arguments Pro and Con." 2006. Center for Genetics and Society. http://www.genet icsandsociety.org/article.php?id=282. Accessed on December 22, 2014.

Rideout, William M., III, Kevin Eggan, and Rudolf Jaenisch. 2001. "Nuclear Cloning and Epigenetic Reprogramming of the Genome." *Science*. 293(5532): 1093–1098.

Robertson, John A. 1994. "The Question of Human Cloning." *Hastings Center Report*. 24(2): 6–14.

Rodriguez, Ramon M., Pablo J. Ross, and Jose B. Cibelli. 2012. "Therapeutic Cloning and Cellular Reprogramming." *Advances in Experimental Medicine and Biology*. 741: 276–289.

"Rooting Hormone for Plant Propagation!" 2014. Olivia's Solutions. http://encyclopedia2.thefreedictionary.com/ Life-Span. Accessed on December 14, 2014.

Rossant, Janet. (1997). "Cloning Human Beings: The Science of Animal Cloning." https://bioethicsarchive.georgetown. edu/nbac/pubs/cloning1/cloning.pdf. Accessed on December 19, 2014, B1–B21.

Sanderson, Katharine. 2008. "Tasmanian Tiger Gene Lives Again." *Nature News*. http://www.nature.com/ news/2008/080520/full/news.2008.841.html. Accessed on December 19, 2014.

Savulescu, Julian. 1999. "Should We Clone Human Beings? Cloning as a Source of Tissue for Transplantation." *Journal of Medical Ethics*. 25(2): 87–95. http://jme.bmj.com/con tent/25/2/87.full.pdf. Accessed on December 20, 2014.

"Science and Engineering Indicators." 2012. Arlington, VA: National Science Foundation. Available online at http:// www.nsf.gov/statistics/seind12/pdf/seind12.pdf. Accessed on December 23, 2014.

Shanks, Peter. 2013. "Should We Be Trying to Bring Extinct Species Back to Life?" Climate Connections. http://

climate-connections.org/2013/04/10/should-we-be-trying-to-bring-extinct-species-back-to-life/. Accessed on December 19, 2014.

Sharples, Tiffany. 2008. "Your Steak—Medium, Rare or Cloned?" Time. http://content.time.com/time/health/article/0,8599,1714146,00.html. Accessed on December 17, 2014.

Sherkow, Jacob S., and Henry T. Greely. 2013. "What If Extinction Is Not Forever?" *Science*. 340(6128): 32–33.

Shikai, Yuriko Mary. 2004. "Don't Be Swept Away by Mass Hysteria: The Benefits of Human Reproductive Cloning and Its Future." *Southwestern University Law Review*. 33(Part 2): 259–284.

Siegel, Andrews. 2013. "Ethics of Stem Cell Research." Stanford Encyclopedia of Philosophy. http://plato.stanford.edu/entries/stem-cells/. Accessed on December 23, 2014.

Smith, Lawrence C., et al. 2000. "Benefits and Problems with Cloning Animals." *Canadian Veterinary Journal*. 41(12): 919–924.

Smith, Lesley J. 2005. "Ian Wilmut: Human Cloner." Discovery Institute. http://www.discovery.org/a/2424. Accessed on December 16, 2014.

"Staff Working Paper 3B: Arguments against 'Reproductive Cloning'." 2002. https://bioethicsarchive.georgetown.edu/pcbe/background/workpaper3b.html. Accessed on December 20, 2014.

"State Initiatives for Stem Cell Research." 2009. Stem Cell Information. http://stemcells.nih.gov/research/pages/stateResearch.aspx. Accessed on December 29, 2014.

"States." 2014. Americans United for Life. http://www.aul.org/your-state/. Accessed on December 29, 2014.

"Stem Cell Information." 2014. National Institutes of Health. http://stemcells.nih.gov/research/pages/newcell_qa.aspx. Accessed on December 28, 2014.

"Stem Cells: Scientific Progress and Future Research Directions." 2001. National Institutes of Health. http://stem cells.nih.gov/staticresources/info/scireport/PDFs/full rptstem.pdf. Accessed on December 11, 2014.

"Summary of State Laws on Human Cloning." 2011. Bioethics Defense Fund. http://bdfund.org/wordpress/wp-content/ uploads/2012/07/CLONINGChart-BDF2011.docx.pdf. Accessed on December 29, 2014.

Tabar, V., et al. 2008. "Therapeutic Cloning in Individual Parkinsonian Mice." *Nature Medicine.* 14(4): 379–381.

Tannert, Christof. 2006. "Thou Shalt Not Clone." *EMBO Reports.* 7(3): 238–240.

"Tell FDA You Don't Want Cloned Meat for Dinner!." 2014. Care2 Petitions. http://www.thepetitionsite.com/takeac tion/954/622/138/. Accessed on December 17, 2014.

Thompson, Bert, and Brad Harrub. 2001. "Human Cloning and Stem-Cell Research—Science's 'Slippery Slope' [Part I]." http://www.apologeticspress.org/ apcontent.aspx?category=7&article=1355. Accessed on December 16, 2014.

Tachibana, Masahito, et al. 2013. "Human Embryonic Stem Cells Derived by Somatic Cell Nuclear Transfer." *Cell.* 153(6): 1228–1238.

Takahashi, Kazutoshi, and Shinya Yamanaka. 2006. "Induction of Pluripotent Stem Cells from Mouse Embryonic and Adult Fibroblast Cultures by Defined Factors." *Cell.* 126(4): 663–676.

Thompson, Tamara. 2015. *Genetically Modified Food.* Farmington Hills, MI: Greenhaven Press.

Till, J. E., and E. A. McCulloch. 1961. "A Direct Measurement of the Radiation Sensitivity of Normal Mouse Bone Marrow Cells." *Radiation Research.* 14: 213–222.

Vawter, Dorothy E. 1993. "Fetal Tissue Transplantation Policy in the United States." *Politics and the Life Sciences.* 12(1): 79–85.

Vial Correa, Juan de Dios. 1997. "Reflections on Cloning." Pontifica Academia pro Vita. http://www.vatican. va/roman_curia/pontifical_academies/acdlife/documents/ rc_pa_acdlife_doc_30091997_clon_en.html#ETHICAL PROBLEMS CONNECTED WITH HUMAN CLONING. Accessed on December 21, 2014.

Volney, Krystal. 2013. "Human Cloning." Wall Street International. http://wsimag.com/science-and-technology/4881-human-cloning. Accessed on December 21, 2014.

Warburton, David. 2015. *Stem Cells, Tissue Engineering, and Regenerative Medicine*. Hackensack, NJ: World Scientific.

Wertz, D. C. 2002. "Embryo and Stem Cell Research in the United States: History and Politics." *Gene Therapy*. 9(11): 674–678.

"What Is Agrobiodiversity." 2004. Food and Agricultural Organization of the United Nations. *ftp://ftp.fao. org/docrep/fao/007/y5609e/y5609e00.pdf*. Accessed on December 13, 2014.

Wheat, Kathryn, and Kirstin Matthews. (2007). "World Human Cloning Policies." http://www.ruf.rice.edu/~neal/ stemcell/World.pdf. Accessed on December 29, 2014.

Winston, Robert M. L. 2009. *The Evolution Revolution*. London: Dorling Kindersley.

Winter, P. C., G. I. Hickey, and H. L. Fletcher. 1998. *Instant Notes in Genetics*. Oxford, UK: Bios Scientific Publishers; New York: Springer.

"With Few Factors, Adult Cells Take on Character of Embryonic Stem Cells." 2006. EurekaAlert! http://www. eurekalert.org/pub_releases/2006-08/cp-wff080906.php. Accessed on December 23, 2014.

Woestendiek, John. 2010. *Dog, Inc.: The Uncanny inside Story of Cloning Man's Best Friend*. New York: Avery.

"Woman Pays $50,000 to Clone Dog." 2012. Animal Planet. http://blogs.discovery.com/daily_treat/2012/01/woman-pays-50000-to-clone-dog.html. Accessed on December 17, 2014.

Wynn, Rebecca. 2014. "Hello Dolly, Hello Dolly: Human Cloning, Ethics and Identity." Pro+Choice Forum. http://www.prochoiceforum.org.uk/ri4.php. Accessed on December 22, 2014.

Young, Emma. 2002. "First Cloned Baby 'Born on 26 December'." New Scientist. http://www.newscientist.com/article/dn3217-first-cloned-baby-born-on-26-december.html#.VIsb5CvF-So. Accessed on December 12, 2014.

Zimmer, Carl. 2013. "Bringing Them back to Life." National Geographic. http://ngm.nationalgeographic.com/2013/04/125-species-revival/zimmer-text. Accessed on December 19, 2014.

de, Höhle Ekain, Spanien

3   Perspectives

Introduction

Cloning is a topic of considerable interest to a wide range of individuals both in the scientific world and among the general public. This chapter provides an opportunity for some of these individuals to express their viewpoints on various aspects of the topic of cloning, ranging from advances in the science and technology of cloning to the arguments for and against various types of cloning.

Should We Clone Dinosaurs?
Sandy Becker

We still can't clone dinosaurs, or even mammoths, or any other extinct animal. Michael Crichton is lucky he wrote Jurassic Park in 1990, before we realized how hard it would be to clone a dinosaur! In 1958 John Gurdon successfully cloned a frog, using the nucleus of a tadpole somatic cell. In 1996 Dolly the sheep was cloned from a mammary cell. (She was named after Dolly Parton, the country western singer.) In 1997 a mouse was cloned, named Cumulina because she was cloned from cumulus cells. As of this writing the list of cloned animals contains 23 entries, some endangered, but none extinct.

Cloning any multicelled animal involves inserting a nucleus from the animal to be cloned into an enucleated egg, which

Japanese researchers announced in 2011 that they would pursue efforts to clone the extinct woolly mammoth, a replica of which is shown here at the Neanderthal Museum in Mettmann, Germany. (Roland Weirauch/epa/Corbis)

will instruct the nucleus to forget it belongs, for example, to a skin cell and adopt the properties of an embryonic cell, being able to divide and differentiate and eventually produce all the cells of the body. This process is called somatic cell nuclear transplantation, or SCNT. At best, it is very inefficient. A great many nuclei must be inserted into a great many eggs before one of them gets the message. It took Ian Wilmut 277 tries to clone Dolly the sheep.

In animals from flies to lizards to humans, the DNA is organized into chromosomes, in pairs containing one from each parent. For example, humans have 23 pairs of chromosomes; mice have 20, dogs have 39, and fruit flies have 4. Each chromosome contains hundreds to thousands of genes, and each nucleus contains not only the chromosomes but also proteins to package the DNA and to instruct each gene to get busy—or to do nothing. "Busy" for a gene means making messenger molecules (RNA) to send out of the nucleus, where they will be used as template for making the many proteins each cell needs. Naked DNA, without this packaging and these proteins, can't carry out the business of running and replicating a cell, let alone the complex choreography of embryonic development. So, to clone a sheep or a mammoth or a dinosaur, you need the complete nucleus.

The fertilized egg, called a zygote, must not only divide, again and again to form a ball of many cells, but must orchestrate the development of all the varied cell types that will eventually make up the newborn animal. Successful cloning, therefore, requires a nucleus from the donor to give up its simple life and resume the complex life of a zygotic nucleus. This is why cloning is so hard!

Cloning a dinosaur would be easier in some respects than cloning a sheep, or even an extinct mammal such as a mammoth. Like frogs or birds or lizards, the embryo wouldn't have to be implanted in a surrogate mom to gestate. But in some ways it would be a lot harder. Where would you find the intact nucleus of a dinosaur cell?

In 1994, a few years after Jurassic Park was published, Scott Woodward and a team of scientists in Utah demonstrated that it just might be possible to get DNA out of dinosaur bones (Woodward et al. 1994). More recently, a multidisciplinary team extracted some identifiable collagen (the main protein component of bone and cartilage) from a specimen of *Tyrannosaurus rex* found in Montana. Their article was published in *Science*, a prestigious, peer-reviewed scientific journal that does not generally publish the work of crackpots (Schweitzer et al. 2007).

But even if you did locate a complete dinosaur nucleus, what would you put it into? In cloning, the recipient egg is just as important as the donor nucleus, as it is the egg cytoplasm that provides the signals to reprogram the nucleus. It is the egg that instructs genes in the nucleus to make proteins needed for an embryo's development rather than whatever they were making when they were, for example, in a skin cell. It is the egg that instructs the donor nucleus to resume cell division.

The cloning scheme described in Jurassic Park involved scavenging scraps of dinosaur DNA from the stomach of a mosquito, preserved in amber, that had feasted on dinosaur blood just before settling on a sappy tree branch and getting stuck. The incomplete dinosaur genome was supplemented with frog DNA and then injected into a crocodile egg to develop.

In 1996 Hendrik Poinar and an international team of researchers published an article in *Science* showing that amber is indeed an excellent preservative for ancient DNA (although probably not good enough to yield an intact nucleus) (Poinar et al. 1996).

Well, what about mammoths? Elephant oocytes could serve as recipients for the donated mammoth DNA, surely a better match than crocodile eggs for dinosaur nuclei! Surely could an intact nucleus be recovered? Probably not. Tissue that is promptly and carefully frozen in liquid nitrogen (about −200°C) when it is freshly dead can be preserved indefinitely, but the permafrost isn't *that* cold, and the natural process of dying and freezing slowly is unlikely to have preserved any intact nuclei for 20,000 years. In 2000, workers at a Massachusetts biotech company, Advanced Cell Technology, cloned an endangered ox-like animal, a gaur. They used cryopreserved skin cells and transplanted them into cow eggs (Vogel 2001). In 2008, Sayaka Wakayama and colleagues from the RIKEN Institute in Japan cloned healthy mice from bodies that had been frozen for 16 years. They found intact nuclei in the chilled bodies and fused them with enucleated eggs. But 16 years in the freezer is one thing; 16,000 years in the permafrost is quite another (Wakayama et al. 2008).

Not to mention the greater challenge of finding a dinosaur cell, millions of years old, with an intact nucleus. Remember that a whole nucleus, not just naked DNA, is inserted into the waiting egg. While scientists in their wildest dreams can hope to find some intact dinosaur DNA preserved somewhere, somehow, the chances of finding an intact nucleus are pretty slim.

And anyway, should we clone an extinct animal, even if we could? It would be a lonely creature, for sure. There would likely be no suitable habitat available. What exactly would we feed it? Better to focus our efforts on cloning endangered animals, some of which are important members of their ecosystems. Of course, then we'd have to preserve those ecosystems, but that's another essay.

References

Poinar, H. N., et al. 1996. "Amino Acid Racemization and the Preservation of Ancient DNA." *Science*. 272(5263): 864–866.

Schweitzer, M. H., et al. 2007. "Analyses of Soft Tissue from *T. rex* Suggest the Presence of Protein." *Science*. 316(5822): 277–280.

Vogel, Gretchen. 2001. "Cloned Gaur a Short-lived Success." *Science*. 291(5503): 409.

Wakayama, S., et al. 2008. "Production of Healthy Cloned Mice from Bodies Frozen at –20°C. for 16 Years." *PNAS*. 105(45): 17318–17322.

Woodward, S. R., et al. 1994. "DNA Sequence from Cretaceous Period Bone Fragments." *Science*. 266(5188): 1229–1232.

Sandy Becker is a biologist at Wesleyan University. She also writes about science.

Should Breed Registries Accept Clones?
Ryan Bell

When Dolly the cloned sheep was born, in 1996, she captured the world's attention. At the time, most people hadn't heard of the Finish-Dorsett breed of sheep, a variety common in Scotland where Dolly was born. But for years to come, Dolly would be the center of an international ethics debate over whether or not a clone is a natural animal. Meanwhile, the clone kingdom grew at a near-annual rate: cow, pig, cat, goat, and rabbit.

Then, on May 28, 2003, the first cloned horse was born in Cremona, Italy. Researchers named her Prometea, after Prometheus, the character in Greek mythology who stole fire from the gods and gave it to humankind. Their hope was that the name would make the clone brave in the face of the prejudice she would encounter in life. Her christening proved to be prophetic. In 2004, the American Quarter Horse Association (AQHA) created a rule prohibiting clones, even though a cloned horse had yet to be born in the United States. The rule was a preemptive measure against a breeding technique most knew about only through the Dolly debate.

In the United States, the Quarter Horse breed is a $100 billion industry. Top-performing horses can win over a million dollars during a career in rodeo or horseracing. When horses retire, they go on to equally lucrative careers as breeding animals. With that much money on the table, it was inevitable that Quarter Horse breeders would try cloning prized horses in the hope of replicating past success. That likelihood became all the more probable when the Texas-based company ViaGen bought exclusive license to the technology patents used by the scientists in Scotland who cloned Dolly. ViaGen has now grown to become the world leader in cloning, with facilities in Texas, Iowa, Canada, Italy, Brazil, and Argentina. It has cloned 300 individual horses, representing a variety of breeds, including 35 individual Quarter Horses.

When Texas horse rancher Jason Abraham learned about cloning, he saw a business opportunity. Abraham worked with a cloning expert, veterinarian Gregg Veneklasen, to clone a stable of champion Quarter Horses. Their aim was to produce and sell the offspring, creating the horse equivalent of a generic product. The horses would be knockoffs of the original, but costing a fraction of the price. For the business plan to work, Abraham and Veneklasen needed AQHA to change its rule against cloning. The reason was that some fields of competition, such as horseracing, require contestants be registered with a breed association. Without AQHA papers, the Abraham-Veneklasen horses couldn't compete and were therefore valueless, no matter their genetic lineages.

In 2012, Abraham and Veneklasen petitioned AQHA to change the cloning rule. They argued that as "identical twins separated by time" clones deserved the same rights of registration afforded to their originals. AQHA denied the request. But Abraham and Veneklasen suspected their proposal hadn't been treated with due process. They filed an antitrust lawsuit against AQHA, alleging that a small number of board and committee members had conspired to kill their proposal. Their reason for doing so, Abraham and Veneklasen alleged, was based on the self-interest of keeping clones out of AQHA so the animals wouldn't undercut the value of existing breeding operations.

The lawsuit was held in a federal courtroom in Amarillo, Texas. The jury found in favor of Abraham and Veneklasen. The judge ordered AQHA to change its registration rules to accept clones. The plaintiffs called the ruling a win against a "good ol' boys" network within the AQHA. The breed organization portrayed it as an injustice and a case of the federal government meddling in the affairs of a member-run organization. National headlines tended to agree with the latter. "Seabiscuit 2? World's Largest Horse Registry Forced to Include Cloned Horses" and "Horse Cloners Try to Force Their Way into the Starting Gate."

The federal court in Amarillo, Texas, agreed to give AQHA a stay so that it could postpone registering clones while it appealed the case to a higher court. In January 2015, the Fifth Circuit Court of Appeals made its ruling in New Orleans, Louisiana. A three-judge panel decided to overturn the district court's ruling. They said that the jury had reached an incorrect verdict in deciding that AQHA had violated Sections One and Two of the Sherman Antitrust Act. Specifically, the appeals court said that Abraham and Veneklasen had not provided sufficient evidence to prove that a faction had formed within AQHA or that it had acted in a way to influence the organization's decision about not registering clones. The reverse ruling canceled the order that AQHA be forced to recognize cloned horses in its breed registry.

In the end, the *Abraham-Veneklasen v. AQHA* lawsuit was not about the science of cloning, but about the business handlings of a breed organization. But the ruling did cement in people's minds the idea that, even though a cloned horse is genetically identical to its original, a copy of a horse is a breed apart.

Ryan Bell is a freelance journalist specialized in writing about the horse industry and international ranch agriculture. He covered the Abraham-Veneklasen v. AQHA *lawsuit in a two-part feature for* Western Horseman. *His reporting has appeared in a number of publications, including* Bloomberg News, Outside, Popular Science, *and* Equus. *His writing has been honored at the Society of American Travel Writers Awards, FOLIO's Eddie and Ozzie Awards, and the American Horse Publications Awards. Ryan holds an MFA in creative nonfiction from The University of Montana.*

Cloning Can Turn Yeast into a Painkiller Factory
Maria Costanzo

Synthetic biology (Synthetic Biology Project, 2015) is an exciting new discipline that has enormous potential to affect our lives by producing useful drugs, foods, and biofuels. Most biologists study molecules or processes that already exist in natural organisms. But synthetic biologists actually design and create new genes and new processes, with results that Mother Nature never dreamed of.

Synthetic biology wouldn't be possible without cloning. When synthetic biologists mix and match genes to engineer new processes, their "parts list" is made up of genes that have been cloned from various organisms. And cloning allows the scientists to modify the genes so they'll be functional in the species that hosts the newly created process.

The following example will show you how synthetic biology could lead to big improvements in the production of an important class of drugs. Opiate drugs (e.g., morphine) are used to control pain, and they are an essential part of modern medicine. Currently, the raw material to make these drugs comes from opium poppies. Just like any other crop, the yield depends on the weather, so the supply isn't completely reliable. And because opiate drugs can be abused, we'd like to have better control over every step of their production—which would be easier if they were produced in a fermentor than in the field.

Dr. Christina Smolke of Stanford University set out to find a better way to produce opiate drugs. She and her group wanted to see whether they could turn the microorganism *Saccharomyces cerevisiae* into a drug factory.

People have harnessed the power of *S. cerevisiae*, also known as baker's yeast, for thousands of years (Saccharomyces Genome Database, 2013). This is a single-celled fungus that has helped bakers and brewers over the centuries. More recently its genetics and molecular and cellular biology have been studied in the lab, making it a great model organism for basic research, which is aimed at understanding the universal processes of life. It is also a very important organism for applied research, which has results that are useful for society—anything from making a beer that tastes better, to the production of medicines or even jet fuel.

In a recent study, Thodey, Galanie, and Smolke (2014) set out to make opiate drugs by yeast by recreating the biochemical pathway for opiate synthesis in yeast. A biochemical pathway is a series of sequential enzymatic reactions, very much like a factory assembly line. The first enzyme takes a precursor chemical and converts it into a product. This product then becomes the precursor for the next enzyme, and so on. Synthetic biology is all about putting together these assembly lines, using enzymes from different organisms as the steps in assembly.

The researchers cloned three genes from opium poppy plants. These genes code for enzymes in the opiate synthesis pathway, called thebaine 6-O-demethylase (T6ODM), codeine O-demethylase (CODM), and codeinone reductase (COR). Of course, simply transforming yeast with a plant gene doesn't do much good; different organisms don't speak exactly the same language when it comes to gene regulation. So the researchers put the poppy genes under the control of efficient yeast transcriptional regulatory sequences such as promoters and terminators and optimized their codons for yeast.

When they had expressed all three enzymes in yeast, the researchers fed the cells with the molecule thebaine, which is a precursor in the pathway, and the engineered cells started producing opiates. However, the assembly line didn't work quite as well as they hoped. It was not very efficient, and it generated the unwanted opiates neomorphine and neopine in addition to the desired products, morphine and codeine.

Thodey and colleagues tinkered with the expression of the genes and were able to improve the efficiency of the pathway, but almost half of the product was still the undesirable neomorphine. To address this problem, they looked even more closely at the details of the pathway.

When morphine synthesis is going right, the chemical product made by the T6ODM enzyme spontaneously rearranges its structure to form a different chemical that COR uses to continue along the pathway. But if COR grabs the T6ODM product before there is time for the rearrangement, the end result of the pathway is neomorphine, which is not useful.

The researchers decided to separate T6ODM and COR into different parts of the cell, to allow more time for this rearrangement. They added a tag to COR that would direct it to the membrane of the endoplasmic reticulum, a subcompartment of the cell, while T6ODM stayed in the cytoplasm. Now it would take longer for the T6ODM product to reach COR, giving it plenty of time to rearrange into codeinone. Sure enough, morphine production went way up.

This result in itself was extremely useful, but the researchers decided to see whether yeast could be even more useful. Semisynthetic opioids such as hydrocodone, oxycodone, and hydromorphone are medically important because they work better in some cases than the natural opiates. Currently, they are produced by chemical modification of the opiates produced by poppies. Could yeast do this job too? Of course!

To accomplish this part of the pathway, Thodey and colleagues tried expressing the enzymes NADP+-dependent morphine dehydrogenase (morA) and NADH-dependent morphinone reductase (morB) from the bacterium *Pseudomonas putida* in yeast, along with the poppy enzymes. Again, the process needed a lot of tweaking, but the end result was a strain that produced both hydrocodone and oxycodone.

Putting together all their results, the researchers were able to construct three yeast strains, each like an assembly line tailored for different products. One assembly line was optimized for codeine and morphine, another for hydromorphone, and one for hydrocodone and oxycodone.

The next steps will be to scale up this process to industrial levels and also to construct yeast strains that carry out the entire process starting from simple sugars, rather than needing to be fed the precursor thebaine. Substituting yeast cultures for opium poppy fields will have a huge global impact that goes far beyond pharmaceutical production.

This story is just one example. The potential of cloning useful genes and engineering them into pathways has only just begun to be realized. Genome sequencing is being done for more and more organisms and that means that we have a growing selection of genes, encoding enzymes with many different abilities, to use in constructing our assembly lines. Look for exciting new advances in synthetic biology in the coming years!

References

Saccharomyces Genome Database. 2013. http://wiki.yeast-genome.org/index.php/What_are_yeast%3F. Accessed on January 28, 2015.

Synthetic Biology Project. 2015. http://www.synbioproject.org/topics/synbio101/. Accessed on January 28, 2015.

Thodey K., S. Galanie, and C. D. Smolke. 2014. "A Microbial Biomanufacturing Platform for Natural and Semisynthetic Opioids." *Nature Chemical Biology*. 10(10): 837–844.

Maria Costanzo is a senior biocurator at the Saccharomyces Genome Database (www.yeastgenome.org), a freely available online database that collects information about the genes, proteins, and genome of the model organism Saccharomyces cerevisiae, or baker's yeast. She has a PhD in cellular and developmental biology and has done basic research on yeast.

Cloning for Liver Transplantation
Gina Hagler

A sufficient quantity of organs for transplant is one of the greatest needs in modern medicine. The transplant donor registry is used to match patients to available organs, but there is a chronic shortage of organs. One potential source of organs in the future would be those created by cloning the cells for that type of organ. The cloning process has been refined to the point that the stem cells required can be created with the use of an unfertilized egg rather than through the use of an embryonic stem cell, but there is concern that there will not be a sufficient supply of eggs for this purpose. It would be far better if stem cells for specific organs could be created without the use of any sort of egg or embryo.

Induced pluripotent stem (IPS) cells are pluripotent stem cells that are created without the use of egg or embryo. With an IPS cell, genes are inserted into a mature cell. The mature cell already has a specific function, but the insertion of the genes reverts the purpose of the cell back to an earlier state, a state in which the purpose of the cell was not yet set. The introduction of the gene will reprogram this induced pluripotent stem cell to perform a new function.

The resulting reprogrammed cell is an exact genetic match to the patient. The risk of rejection of IPS cell organs is minimal, because all of the changes to the cell took place in one of the patient's own cells. It would seem an ideal solution to create genetically matched organs in this way, but for the fact that there are some concerns about the behavior of IPS cells over the long term. Before this can be investigated, human organs need to be created with the use of IPS cells.

Discover magazine reported on July 3, 2013 (Lang 2013), that a team of Japanese researchers at Yokohama City University had recently used IPS cells to create a human liver. This organ, the first produced with IPS cells, was fully functioning. The process used to create the liver involved the creation of reprogrammed human cells to create liver buds. These bud cells were transplanted into mice and their growth was tracked. The process used to develop, track, and test the organ resulting from these cells will serve as a model for organ creation using IPS cells in the future.

In the Japanese study, the researchers began with human skin cells, a type of cell commonly used in cloned stem

cells. They reversed these cells to IPS cells before inserting the material for liver creation. The material for a liver
alone was not enough because the liver requires a blood
supply as well. Stem cells that line blood vessels, along with
those that form tissue, were added to the dish with the cells.
Proto-organs that could be seen without a microscope were
produced within 48 hours. These proto-organs were the
liver buds.

Before the liver buds were implanted into easily accessible
sites on mice, they were marked with fluorescent proteins.
These marked proteins made it possible to track the development of blood vessels. The organ developed a vascular system
almost immediately. Other indicators that the organ being developed was indeed a human liver were also noted. To prove
that the organ was a true human liver, researchers gave the mice
drugs that cannot be metabolized by a mouse liver. The transplanted livers broke these drugs into the same components as
a human liver does.

One finding of interest was that the transplanted liver resulting from the IPS process need not be located near the existing
liver. In fact, future plans call for liver buds to be made small
enough that they can be injected directly into the bloodstream
of the mice being studied. One drawback to this method of
organ production is that currently the livers that can be grown
in this manner have the capacity of about 30 percent of a functioning adult liver. However, these organs could serve as a sup-
plement for someone awaiting transplant. They could also be
used in critically ill children and infants with catastrophic liver
damage who would otherwise die.

The promise of IPS cell organs is great. The research is in its
infancy. More will be learned through additional experimentation and refinement of the process. There is an unlimited supply of these cells that are a perfect match to the patient. Early
experiments with cell cloning have been performed using genetic material from skin cells. It is logical that IPS cells can be
created from the same genetic material. If the organs created

with this method prove to be effective for an acceptable period of time, it will mean that the organs produced in this way are more viable for transplantation into a patient than a donor organ. Patients will not have to wait until an organ that matches their needs becomes available. They will be able to create an organ that is a perfect match when it is needed without the need for a waiting list or a donor.

Most researchers believe that the use of this technology in humans is at least a decade away. What an exciting decade that will be.

Reference

Lang, Becky. 2013. "First Functioning Human Organ Made of Induced Stem Cells—D-brief." *Discover*. http://blogs. discovermagazine.com/d-brief/2013/07/03/human- liver-is-first-functioning-organ-made-of-induced-plu ripotent-stem-cells/#.VNjVDsZcbuU. Accessed on February 8, 2015.

Gina Hagler is a freelance writer and published author who covers science, technology, health, climate change, and species survival. She is a member of the National Association of Science Writers (NASW), the American Society of Journalists and Authors (ASJA), and the Society of Environmental Journalists (SEJ). You'll find more of her work at www.ginahagler.com.

Extinct Species Should Not Be Resurrected
Phill Jones

Scientists are investigating how cloning methods and recombinant DNA technology can be used to bring extinct species back into the world. This endeavor is called *de-extinction*. Although the idea sounds like the motion picture *Jurassic Park*, nobody seriously suggests that dinosaurs should be resurrected. De-extinction efforts focus on species that became extinct within the last several tens of thousands of years. The remains of these species offer the best chance for recovering intact cells or enough DNA to analyze and reconstruct a genome.

For example, researchers from North-Eastern Federal University in Siberia are recovering frozen tissue of woolly mammoths. Scientists at the Sooam Biotech Research Foundation in South Korea analyze tissue samples. One possible method to resurrect the woolly mammoth is to remove the nucleus from an elephant egg cell and insert a nucleus isolated from a cell found in frozen mammoth tissue. After stimulating the egg cell to divide, the cells are placed in the uterus of an elephant. Theoretically, the elephant gives birth to a baby woolly mammoth.

De-extinction is appealing, especially because humans have caused the extinction of so many species. Yet Paul R. Ehrlich, a professor of population studies at Stanford University, characterized the resurrection of extinct species as "a fascinating, but dumb idea" (Ehrlich 2014). At least four arguments support the position that de-extinction should not be pursued.

De-Extinction Efforts Cause Animal Suffering

A wild goat called a bucardo lived in the mountain range between France and Spain for thousands of years. During the last several centuries, hunters decimated the bucardo population until one female bucardo named Celia remained in 1999. The bucardo species became extinct after a tree fell on Celia. Before she died, a scientist scraped a small piece of the skin and froze the tissue.

A team of researchers wanted to resurrect the bucardo species. They isolated nuclei from Celia's skin scrapings, removed nuclei from goat eggs, inserted Celia's nuclei into the goat eggs, and implanted the eggs in female goats. Fifty-seven implantations produced seven pregnant goats; six terminated in miscarriages. After a cesarean section delivery, Celia's clone struggled to breathe. The clone died 10 minutes later. A study of the clone revealed that a huge piece of solid tissue had grown from one of her lungs.

In hindsight, this failure is not surprising. Animal cloning experiments show that identical genomes do not ensure that two animals will be identical. In addition to nucleotide sequences, epigenetic factors regulate gene expression that is vital for development and survival of offspring. Cloning efficiencies are very low. "[T]he process of pregnancy and birth may be repeated hundreds or thousands of times," says environmental journalist Zion Lights, "with high levels of complications such as abortions, deformed results, and death on delivery amongst the cloned species" (Lights 2013).

Animal clones that survive may suffer from many health problems, such as heart and lung disorders, high rates of tumor growth, and increased predisposition to infections. "De-extinction will not be possible without violating any reasonable standard of humane treatment," says Emory University neuroscientist and animal behavior expert Dr. Lori Marino. "For that reason alone it is unethical" (Marino 2013).

Resurrected Animals May Have No Place to Live

Humans have altered many habitats by deforestation, pollution, illegal hunting, human-driven climate change, disruption of natural waterways, and other activities. Consequently, an ecosystem for a de-extinct species may no longer exist.

Simply placing resurrected animals in the wild would present problems. For example, they would compete with current species for food. The de-extinct animals could even hasten the death of endangered animals. On the other hand, a resurrected species that failed to adapt to a new habitat would become extinct again. Glenn Albrecht, director of the Institute for Social Sustainability at Australia's Murdoch University, warns about de-extinction. "Without an environment to put re-created species back into," he says, "the whole exercise is futile and a gross waste of money" (Zimmer 2013).

Resurrected Animals Could Resurrect Diseases

The body of a resurrected animal could offer a new environment for the development or transfer of disease-causing microorganisms. "What if we bring back something that is actually an excellent vector for diseases, for example, that could affect livestock or other species or ourselves?" asks Dr. Axel Moehrenschlager, Head of the Centre for Conservation & Research at the Calgary Zoo (CBC News staff 2014).

Then there is the retrovirus problem. Retroviruses can insert their genome into the DNA of an infected animal cell. Later, the inserted viral DNA can activate and cause the formation of viruses. The cells of a resurrected animal could include dormant retroviral DNA that activates to create a deadly pathogen that became extinct long ago.

De-Extinction Efforts Distract from the Current Mass Extinction Problem

Many scientists voice concern that the notion of de-extinction diverts attention from a critical problem: At this time, species are becoming extinct at a rate of at least 1,000 times the natural rate. By the end of this century, the world may lose half of all species due to climate change, overhunting, pollution, and habitat destruction (Biello 2013). "From an ethical point of view, why should the world's scientific resources be focused on reviving species for which there are so many unknowns rather than on saving any of the millions of species facing extinction in the next few years?," asks Lori Marino. "What is the validity of promoting the revival of mammoths, for instance, when Asian and African elephants will be lost by 2020?" (Marino 2013).

Wildlife author Paul Guernsey argues that cloning will never provide a solution to the problem of evervanishing species. "If we truly want to keep species from going extinct, the only way we can be successful is to stop destroying them, and to conserve their habitat," he says. "Everything else is just science fiction" (Guernsey 2011).

References

Biello, David. 2013. "Will We Kill off Today's Animals If We Revive Extinct Ones?" *Scientific American*. http://www.sci entificamerican.com. Accessed on December 5, 2014.

CBC News. CBC News staff. 2014. "Species De-Extinction Plagued By 'Looming Questions,' Expert Says." http:// www.cbc.ca. Accessed on December 11, 2014.

Ehrlich, Paul R. 2014. "The Case against De-Extinction: It's a Fascinating but Dumb Idea." Yale Environment 360. http://e360.yale.edu. Accessed on December 12, 2014.

Guernsey, Paul. 2011. "Why Cloning Won't Save Endangered Animals." All about Wildlife. http://www.allaboutwildlife .com. Accessed on December 8, 2014.

Lights, Zion. 2013. "Should Cloning Be Used to Bring Back Extinct Species?" Permaculture. http://www.permaculture .co.uk. Accessed on December 9, 2014.

Marino, Lori. 2013. "Four Reasons Why We Should Oppose 'De-Extinction'." The Kimmela Center for Animal Advocacy, Inc. http://www.kimmela.org. Accessed on December 9, 2014.

Zimmer, Carl. 2013. "Bringing Them back to Life." National Geographic. http://ngm.nationalgeographic.com. Accessed on December 7, 2014.

Phillip Jones, PhD, JD, writes articles and books in the areas of general science, agricultural biotechnology, forensic science, medicine, history, and law.

Animal Cloning
Yoo Jung Kim

In 1952, Robert Briggs and Thomas J. King created the first frog clones by transferring the nuclei from early-stage frost embryos into enucleated eggs—a process called nuclear transfer. This discovery provided the technical foundation that would be adapted for other species, such as Dolly the Sheep. Dolly was the first successful mammalian clone in 1996 through somatic nuclear transfer (SCNT), a process in which genetic material is extracted from developed cells of the body, rather than embryos. Since Dolly, scientists have successfully cloned rhesus monkeys, mice, cows, goats, and pigs—among others (University of Utah Health Sciences 2014a).

Pros of Animal Cloning

Animal cloning has tremendous implications in agriculture, pharmaceutical research, and animal breeding. For example, cloning can be used in animal husbandry to create breeding animals that express a set of desired hereditary traits in agriculture, including higher milk and meat production in livestock. In 2008, to address the growing concern from consumers over potential food produced from animal cloning, the U.S. Food and Drug Administration concluded that the "meat and milk from cow, pig, and goat clones and the offspring of animal clones are as safe as food we eat every day" (U.S. Food and Drug Administration 2014).

Animal cloning could also help researchers maintain genetic mutations within research animals. In China, the Beijing Genomics Institute, one of the biggest DNA sequencing company in the world, is currently cloning pigs at an industrial scale of about 500 animals a year. The clones are created from laboratory pigs that carry genetic mutations—such as those associated with inactive growth genes and susceptibility with Alzheimer's diseases—which enable researchers to mimic human diseases in a model animal that resembles humans in physiology and size. The cloning process expands the number of animals that can be tested, thereby allowing scientists to better test pharmaceutical compounds for safety and efficacy (Shukman 2014).

Furthermore, scientists may also be able to use animal cloning to revive extinct species and to preserve endangered animals. For instance, in 2009, scientists were able to reanimate the Pyrenean ibex—which had been extinct since 2000—by transferring the nuclei derived from preserved ibex tissue into enucleated goat embryos. Unfortunately, the ibex that was born succumbed to a lung defect merely seven minutes after its birth, but the process gave researchers hope that the "de-extinction" of species long gone, such as the woolly mammoth, could soon become a reality (Gannon 2013).

Some companies have even marketed the animal cloning for personal or leisurely pursuits. For example, the second horse in the world to be cloned was Pieraz, a multiple world champion in endurance racing. Pieraz had been castrated at an early age and was unable to reproduce naturally, so Pieraz's clone would theoretically pass on Pieraz's genes to produce progeny with similar high-performance characteristics. Horse cloning companies have sprung up to help equine owners to expand the breeding capacity of their top-performing horses and preserve their animals' genes as insurance against loss or injury (Coghlan 2005). Likewise, a market for animal cloning has opened for dog owners. Hwang Woo-suk—a South Korean scientist accused of fabricating his results in a highly profiled stem cell research paper published in *Nature*—founded Sooam Biotech Research Foundation, which has the capacity to produce 150 to 200 clones a year for those willing to pay $100,000 to clone their beloved dogs (Dean 2014).

Controversies in Animal Cloning

Breeding registries have been slow to accept cloning over fears that owners will choose to clone animals for a limited set of desirable traits. For example, the American Quarter Horse Association (AQHA) has refused to register cloned horses, stating that cloning "has the potential to intensify the narrowing of the gene pool," thereby resulting in genetic diseases that would negatively impact the breed's future well-being (American Quarter Horse Association 2015).

On an even graver note, ethicists and animal rights activists have decried the process over moral and ethical considerations.

For instance, cloning requires that technicians operate on living animals to extract eggs and to implant the resulting embryos into surrogate mothers. However, cloning animals from somatic cell nuclear transfer is highly inefficient, with a success rate ranging from 0.1 to 3 percent. Furthermore, many of the host mothers that carry the fetus suffer from high rates of spontaneous abortions and stressful deliveries, and 30 percent of the clones that are delivered successfully suffer from "large offspring syndrome," a condition that results in enlarged internal organs, underdeveloped lungs, and a host of other health problems including immune deficiencies, diabetes, and brain abnormalities (University of Utah Health Sciences 2014b).

Conclusion

Animal cloning offers many possibilities, including more efficient animal husbandry, better pharmaceutical research, and the potential to protect endangered species and revive extinct animals. However, animal cloning is not without its controversies, chief among them being the reduction of genetic diversity and undue animal suffering because of current inefficiencies in the cloning process. Academic scientists and biotechnology companies must weigh these concerns carefully, lest we forsake the well-being of other animals for the sake of human progress.

References

American Quarter Horse Association. 2015. "American Quarter Horse Association Position Regarding: Abraham & Veneklasen Joint Venture v. American Quarter Horse Association." www.aqha.com/AQHA-Cloning-Law suit-Resources/AQHA-Cloning-Position.aspx. Accessed on January 31, 2015.

Coghlan, Andy. 2005. "First Clone of Champion Racehorse Revealed." *New Scientist*. www.newscientist.com/article/

dn7265-first-clone-of-champion-racehorse-revealed.html#. VM6kM1pCeOI. Accessed on December 30, 2014.

Dean, Josh. 2014. "For $100,000, You Can Clone Your Dog." *Bloomberg Business*. www.bloomberg. com/bw/articles/2014-10-22/koreas-sooam-bio-tech-is-the-worlds-first-animal-cloning-factory. Accessed January 10, 2015.

Gannon, Megan. 2013. "Reviving the Woolly Mammoth: Will De-Extinction Become Reality?" Livescience. www.li-vescience.com/27939-reviving-extinct-animals-mammoths. html. Accessed on January 31, 2015.

Shukman, David. 2014. "China Cloning on an 'Industrial Scale'." BBC News. www.bbc.com/news/science-environ-ment-25576718. Accessed January 10, 2015.

U.S. Food and Drug Administration. 2014. "Animal Clon-ing." https://mail.google.com/mail/u/0/#search/david+new ton+cloning/149f21568a32e26d?projector=1 Accessed on January 16, 2015.

University of Utah Health Sciences. 2014a. "The History of Cloning." Learn.Genetics. Genetics Science Learning Cen-ter. http://learn.genetics.utah.edu/content/cloning/clone-zone/. Accessed on January 20, 2015.

University of Utah Health Sciences. 2014b. "What Are the Risks of Cloning?" Learn.Genetics. Genetics Science Learn-ing Center. http://learn.genetics.utah.edu/content/cloning/cloningrisks/. Accessed on January 20, 2015.

Yoo Jung Kim has conducted research at the National Cancer Institute, Dartmouth Geisel School of Medicine, and the University of Washington Medicine Institute of Stem Cell and Regenerative Medicine, and she is currently a postbaccalaureate research fellow at the National Human Genome Research Institute. She is also coauthoring a book called What Every College Science Student Should Know, *to be published by the University of Chicago Press in 2016.*

Cloning for Species Salvation: Why Not?
Pasqualino Loi and Grazyna Ptak

The production of a normal individual (Dolly the sheep) following the "fertilization" of an oocyte depleted of its endogenous chromosomes with a somatic cell (a procedure defined as "cloning") brought a revolution in animal reproduction. Cloning empowers us to "copy and paste" a date genotype, making in theory infinite numbers of identical animals through an "asexual" process. It is this property that leads scientists to consider cloning as a tool to multiply animals in the brink of extinction (Jabr 2013).

The question is urged by the dramatic reduction of animal biodiversity taking place on the planet. Twenty-three different animals have been cloned so far, ranging from amphibians and fish to insects and mammals, demonstrating the fact that cloning might be applied to the majority of species living on the planet.

Cloning involves the collection of somatic cells from easily accessible sources, skin biopsy, blood, and the transfer of nuclei (by direct injection or cell fusion) into an enucleated oocyte. The enucleated oocyte is artificially activated to start the development, and the resulting embryo is transplanted into a foster mother that will deliver a genetic copy of the nuclei donor. The entire procedure is easily manageable in domestic animals, such as cattle, pig, and sheep, but presents several hurdles in case of rare and endangered animals.

The first obstacle is the availability of oocyte donors. Although somatic cells can be easily harvested (a drop of blood contains tens of thousands of nucleated cells), the problem is the oocyte source. The number of females in an endangered species is too limited, and we know very little about their reproduction. The problem is normally overcome by collecting oocytes for a domestic species genetically close to the one to be cloned, a procedure defined as interspecific cloning.

Interspecific cloning works fine when the rare and domestic species are genetically close, as it has been shown for the mouflon cloned using sheep oocytes and foster mothers, but poses the following problems in case of significant genetic distance between nuclei and oocyte donor:

Incompatibility between genomic and mitochondrial DNA. In interspecific cloning, the nuclear genome comes from the rare or wild animal whereas the mitochondria—with their own genome—from the oocyte donor. Mitochondria are the energy-producing factories in the cells and embryo, and to better preserve the integrity of their genome, crucial mitochondrial genes are transferred into the nuclei, which make possible a more accurate replication of DNA. Hence, a crosstalk between nuclei and mitochondria is essential for proper energy production. When the genetic distance for the species to be cloned and the oocyte donor increases though, it might be that mitochondrial and nuclear genome do not functionally interact, leading to developmental arrest.

Activation of the embryonic genome. The embryonic genome (zygotic) established at fertilization by the fusion of the male and female chromosomes remains inactive during the first embryonic cleavages that are regulated instead by the translation of messenger RNA accumulated during oocyte maturation. At different stages of embryonic development (generally four days after fertilization), oocyte factor(s) triggers the transcriptional machinery of the embryonic genome, which regulates its own development.

Again, when the genetic distance between nuclei and oocyte donors increases, there are realistic chances that the oocyte factors fail to activate the incoming genome, leading to early embryonic arrest.

Current State of the Art

Are we ready to clone rare and endangered animals? Not yet. Cloning efficiency in terms of the number of animals born is too low (1–5%). The low efficiency is due to the incomplete "reprogramming" of the somatic nucleus by the oocyte. In other words, the oocyte fails to erase (reprogram) the modifications of the genome (called epigenetic) established during cellular differentiation, and as a result the majority of the clones die during development, or after birth.

In addition, the incomplete reprogramming that occurs during interspecific cloning adds the two biological roadblocks above. Hence, it is unrealistic and expensive to apply interspecific cloning using the current state of art.

My View

Given that further work is required to make use of interspecific cloning a reliable procedure, my view is that we should "buy" time by establishing biobanks of somatic cells for threatened animals. The recent discoveries that dry storage does not jeopardize the "clonability" of the cells suggests that these biobanks might be established and maintained lyophilized at very low cost, comparing to the standard cryopreservation (Loi et al. 2008). Meanwhile, we can work out technical solutions to the following three problems:

Improving nuclear reprogramming: After 18 years since the first clones, where no major breakthroughs were recorded, recent work by other groups (Matoba et al. 2014) and my own laboratory (unpublished) allows us to foresee realistic possibilities for ameliorating nuclear reprogramming of somatic cells. The trick is to provide the somatic cell nucleus a "format" easily readable by the oocyte, which ultimately will erase all differentiation marks. It is hence realistic to assume that within the next 10 years we can count on reliable cloning efficiencies, in mammals at least.

Incompatibility between nuclear and mitochondrial DNA: No studies have been undertaken to find solutions for the mitochondrial/nuclear genome incompatibility in interspecific cloning, but a possible approach might be to inject mitochondria of the rare species to be cloned into the oocyte. The resulting embryo will be "heteroplasmic," that is, having mitochondria from both the oocyte and nuclear donors, hence, with better changes to develop. Techniques for mitochondrial injections in oocytes have been established in the 1970s and work just fine.

Activation of the zygotic genome: Our knowledge on the earliest factors activating the zygotic genome are scarce but progressing. Once we have identified the very early factors, all we need is to identify the sequence of the specific gene(s) in the animal to be cloned and inject them (messenger RNA/protein) along with the nucleus into the oocyte, making sure it will be expressed at the right time of zygotic activation. In this way, the genome of the somatic cell injected into the donated oocyte will be transcribed and will develop to term. Of all the biological constraints affecting cloning, this last one is not explored at all.

A realistic timetable to find solutions to all hurdles hampering interspecific cloning might be 10–20 years. There is problem though; let's image hundreds of cloned embryos will be produced: where shall we transfer them for development to term?

We just saw that genetically close females can provide oocytes for cloning, but it is out of discussion that the resulting embryo—with a genetic background belonging to another species—can be transferred to the oocyte donors. Among the barriers built by nature to preserve species integrity, uterine receptivity is the most stringent one: an embryo from a different species will be invariably targeted and destroyed. Are there any solutions? The answer is yes, but a little embryological background is required to catch up. The preimplantation embryo is a hollow ball made of two differentially committed cells: trophoblastic cells (TRs), which will give rise to the placenta, and inner mass cells (ICMs), that will generate the fetus proper. The TR cells establish the contact with uterine lining, inducing the uterine receptivity in case of syngenetic embryo, or the rejection in case of heterogenetic embryo.

A possible solution might be the following. It is likely the oocyte donor would act as a foster mother too (remember:

the length of pregnancy is similar or close to the animal to be cloned). Having a normal, fertilized and a cloned embryo, we can isolate by microsurgery the ICM of the cloned embryo and transplant it into the fertilized one of the oocyte donor, whose ICM has been discarded. The reconstructed embryo is finally transplanted into the oocyte donor womb. Being the contact established by the syngenetic TR cells, implant and pregnancy should progress normally. Data gathered in the 1980s indicate that the approach is feasible.

To conclude, cloning is a unique tool capable of multiplying rare animals in a very short time. Its application, however, can be ensured only if proper investments in basic research are provided.

References

Jabr, Ferris. 2013. "Will Cloning Ever Save Endangered Animals?" Scientific American. http://www.scientificamerican.com/article/cloning-and-conservation. Accessed on February 2, 2015.

Loi, Pasqualino, et al. 2008. "Freeze-Dried Somatic Cells Direct Embryonic Development after Nuclear Transfer." *PLoS ONE*. 3(8): e2978.

Matoba, Shogo, et al. 2014. "Embryonic Development Following Somatic Cell Nuclear Transfer Impeded by Persisting Histone Methylation." *Cell*. 159(4): 884–895.

Pasqualino Loi and Grazyna Ptak are embryologists and lecturers at the Faculty of Veterinary Medicine, University of Teramo, Teramo, Italy.

Animal Welfare Concerns about Cloning
Crystal Miller-Spiegel

Animal cloning is often described in a manner that implies it is merely a faster form of selective breeding or assisted reproduction. Yet it is wholly unnatural, with each step of the process involving laboratory manipulations—from the collection of oocytes to an animal's birth. Though animal cloning experiments have been conducted for decades and some have been touted as success stories, they remain largely inefficient and experimental at best.

Animals are used in cloning experiments for a variety of reasons, including to "improve" food or fiber production; to obtain genetically modified organs for transplantation and pharmaceuticals; to conduct biomedical research; and to reproduce endangered or extinct animals, beloved pets, or otherwise valuable animals.

Some researchers have even attempted to patent not just cloning methods but cloned animals themselves as new "inventions." For example, scientists whose cloning experiments resulted in Dolly the sheep attempted to obtain a U.S. patent for animals (specifically cattle, goats, pigs, and sheep) born through their cloning method. Those claims on cloned animals, however, were denied (U.S. Court of Appeals for the Federal Circuit 2014, 7-12).

Cloning's dangerous effect on animals and their well-being is rarely discussed, except to make a case against human cloning or the consumption of products from cloned animals. It is most often described as "inefficient," but what that really translates into is animal suffering and death.

Animal cloning stands out as a particularly concerning experimental use of animals primarily because (1) the cloning process can involve large numbers of animals, far more than breeding; (2) animals suffer and die disproportionately as a result of cloning experiments; and (3) cloning experiments are largely unregulated under the Animal Welfare Act (AWF), because the animals and/or the purpose are not covered by the act.

Cloning just one animal requires various procedures involving several animals. To collect eggs to produce cloned embryos, female animals have their eggs surgically removed or flushed during hormone-induced superovulations—or eggs are obtained after they have been killed. The nuclei are removed from the eggs. Cells from other animals, which may have been killed to collect the cells or which died naturally, are fused into the eggs to produce a clone embryo. These embryos (and usually large numbers of them) are often surgically implanted into so-called surrogate or recipient females, many of whom suffer miscarriages. For example, in bovine cloning only about 5 percent of pregnancies are maintained full term (Maiorka et al. 2015). If cloned embryos survive gestation, surrogate animals often undergo Caesarean section, because cloned neonates can be larger and would complicate the birth. The females used as involuntary "donors" and "surrogates" receive no benefit (e.g., biological fitness, in terms of contributing their own genes to the next generation) and are actually harmed as mere reproductive machines.

Cloned animals that survive birth often struggle, and many die shortly afterward. For example, in a study of 849 cloned piglets that were stillborn or died before weaning, it was found that approximately 50 percent of cloned piglets died during their first month after birth (Schmidt, Winther, and Callesen 2014, 107–108), and depending upon the breed, that rate can be as high as 73 percent. "These results show that pig cloning results in a considerable loss of piglets. . . . [The] general finding is that the problem is related to the cloning technique as such" (Schmidt et al. 2014, 107–108). The piglets died due to various complications such as malformations, arthritis, septicemia, enteritis, and pericarditis. Bovine cloning shows similarly grim results: in a study of cloned calves that died as neonates, gross deformities were documented including heart, lung, and liver abnormalities (Maiorka et al. 2015). The researchers involved directly attributed the abnormal vascular functions to the cloning process.

Much of what we know about cloning experiments and their effects on animals comes from published papers. However, these papers likely represent only a tiny fraction of cloning experiments, and so the outcomes are unclear. In agricultural research, especially, there is little to no incentive to medically treat and recover animals who are otherwise destined to be slaughtered.

Cloned animals also may or may not exhibit the desired genotype (i.e., carry a certain gene) or phenotype (i.e., exhibit a certain appearance or behavior), and, if not, they are considered to be byproducts, discarded with little to no value. This is especially true for cloned pet dogs, whom clients expect to look and behave like the original dog. The fate of such animals is unclear. Then, there's demand, and what happens if cloned animals or their products fail to be marketable? The company responsible for cloning the infamous Dolly later killed thousands of transgenic sheep that were used to produce milk containing a drug for humans, after the drug trial funding was cut short (Anonymous 2008).

In the United States, animals that are used in agricultural (i.e., to improve food or fiber) or pet cloning–related experiments are excluded from federal animal welfare oversight afforded by the AWA. Mice and rats, which are the mammals most commonly used in laboratory experiments (and commonly used in cloning and genetic engineering experiments), are also excluded from the act. Therefore, it is difficult to assure that basic animal welfare considerations (including ethical justification and a veterinary treatment protocol) and principles are being implemented and to assess the number of animals involved in cloning experiments and their fate.

For 14 years, Gallup public opinion polls consistently show that about 60 percent of the U.S. public believes that animal cloning is morally wrong (Gallup News Service, 2014). Values can and should contribute to public policy, and animal welfare is too often ignored in scientific articles or flashy news stories. The law is failing these animals. Even if animal cloning can be accomplished, the question remains "Is it worth the cost to animals?"

References

Anonymous. 2003. "Dolly Creators Begin Mass Slaughter." The Scotsman. http://www.scotsman.com/news/scotland/ top-stories/dolly-creators-begin-mass-slaughter-1-655826. Accessed on February 15, 2015.

Gallup News Service. 2014. "Gallup Poll Social Series: Values and Beliefs. Q18.I." http://www.gallup.com/file/ poll/170798/Moral_Acceptability_140530.pdf. Accessed February 12, 2015.

Maiorka, P. C., et al. 2015. "Vascular Alterations Underlie Developmental Problems Manifested in Cloned Cattle before or after Birth." PLoS ONE. http://journals.plos. org/plosone/article?id=10.1371/journal.pone.0106663. Accessed on February 14, 2015.

Schmidt, M., K. D. Winther, and H. Callesen. 2014. "Postmortem Findings in Cloned and Transgenic Piglets Dead before Weaning." *Reproduction Fertility and Development.* 27:107–108.

U.S. Court of Appeals for the Second Court. 2014 "RE Roslin Institute (Edinburgh)." http://www.cafc.us-courts.gov/images/stories/opinions-orders/13-1407. Opinion.5-6-2014.1.PDF. Accessed on February 14, 2015.

Crystal Miller-Spiegel holds a MS degree from the Center for Animals and Public Policy at the Tufts University Cummings School of Veterinary Medicine. She is a senior policy analyst for the American Anti-Vivisection Society and cofounder of Down to Earth Farm—Sanctuary for Animals in Bucks County, Pennsylvania.

Can Cloning Save the Honeybee?
Jeremy Summers

For centuries, technological innovation has allowed humans to make great strides in fighting disease, increasing the quality and longevity of life, and addressing other global issues. In the digital age, techniques once thought of as confined to the realms of science fiction—such as cloning—are now tools in the arsenal for scientists to break new ground in solving problems, especially those caused by humans.

Cloning techniques have already been used to treat serious diseases such as cancer and heart disease and to improve human life expectancy and the quality of crops we eat. So what other problems can scientists solve using cloning techniques?

You are probably aware that honeybees play a vital role in food production. But you might not know just how important they really are. According to the U.S. Department of Agriculture, bees aid in the pollination of crops that are worth more than $200 billion each year. These crops, which include cashews, broccoli, cabbage, watermelons, cucumber, strawberries, macadamia nuts, and almonds, simply would not be able to be produced on a large enough scale to meet demand without the aid of the honeybee (Walsh 2013).

In recent years, however, scientists have noticed large numbers of bees dying—and no one knows exactly why. In 2006, a number of commercial beekeepers noticed that unusually high numbers of their adult worker honeybees fled the hive and ended up dead somewhere else. This, of course, led to a rapid loss of the colony. The most alarming fact was that this was not simply affecting a few farms or even certain geographical regions—it was occurring all over the country.

This phenomenon came to be described as colony collapse disorder (CCD), a sort of blanket term to describe large-scale deaths of honeybee populations throughout the United States and the rest of the world. No one knows for sure what causes CCD, but it is most likely due to a range of causes, including a parasitic mite called Varroa destructor often found in decimated colonies, several viruses and bacteria, as well as increased use of pesticides.

Environmentalists have pointed to pesticides—specifically powerful neonicotinoids—as the real culprit behind CCD, and numerous scientific studies support this theory. Because these pesticides are aimed at other insects, they are not directly killing the bees. They are, however, exposing bees to sublethal dosages of poisons through the nectar and pollen honeybees regularly come into contact with. This exposure, some argue, is interfering with their "internal radar" and disrupting their instinctive ability to gather pollen and return safely to the hive.

The jury is still out on what exactly is causing CCD, but there is no argument over its adverse effects. On average, commercial beekeepers see a 10 to 15 percent loss of their colony because of a number of causes. But since 2006, when CCD was first reported, beekeepers have seen mortality rates for commercial populations rise as much as 28 to 33 percent. This increase translates to some staggering numbers. Since 2006, it is estimated that more than 10 million beehives, each with a commercial worth of about $200, have been lost, costing beekeepers alone more than $2 billion. Additionally, the number of honeybee colonies in the United States, which exceeded 6 million in the 1950s, is now down to 2.5 million (Walsh 2013).

So what can scientists do to reverse the effects of CCD and restore bee populations around the globe? The solution may require some innovative thinking. This is precisely where cloning techniques might provide invaluable tools for scientists.

While many opponents of cloning object on religious or ethical grounds, there is a much more significant roadblock to using cloning techniques to help save the honeybee. If scientists simply clone honeybees to increase population numbers, there will be a significant lack of biodiversity, or genetic variation of the species, within these populations. This diversity is an integral part of a healthy and sustainable ecosystem. It also provides several direct benefits to humans. Greater biodiversity leads to increased overall crop yield; naturally pest-resistant crops; stronger medicine; and a plentiful array of woods used for paper, construction, and fuels.

However, this might not be as big a roadblock as scientists once thought, thanks to a better understanding of the unique genetic makeup of honeybees. Most animals have two sets of chromosome—one that they inherit from their father and one that they inherit from their mother. Humans, for example, have 46 chromosomes, with 23 coming from each parent. Bees are different, though. Females, workers, and queens have 32 chromosomes, 16 of which come from the queen's eggs and 16 from the drone's sperm. Drones, though, hatch from unfertilized eggs and therefore have only the 16 chromosomes that were in the egg. Because the queen is genetically composed of 32 chromosomes and can pass on only 16 to her egg, each egg is a unique and different collection of her chromosomes. Drones, by contrast, start with only 16 chromosomes, because they hatch from unfertilized eggs, making each sperm from a drone genetically identical. This is, essentially, a naturally occurring cloning process, making each drone a clone.

To offset this problem and increase biodiversity within the colony, queens regularly mate with 10 to 20 different partners. This creates different subfamilies within the colony, each with the same mother but a different father, meaning that workers that belong to the same subfamily are related by 75 percent, a significant increase over the 50 percent relation that connects the parents and offspring or siblings in almost all other animals.

This phenomenon not only helps explain the honeybees' unique social structure but also perhaps positions them as an ideal candidate for use in exploring the benefits of cloning techniques to address a very real problem. Theoretically, scientists might be able to recreate the natural process of bee reproduction. Rather than cloning a single bee multiple times over to increase the population, scientists might be able to clone enough nondrones to produce a group within the population that still allows the queen enough diversity within potential mates and then let the bees do the rest. Such a process would also allow for genetic mutations to continue to occur, providing even more biodiversity.

Of course, this process is all still largely theoretical. But this is nevertheless an excellent example of how cloning might be used not as a blunt instrument, like a hammer, but rather as a fine instrument, like a scalpel, to help solve nature's most pressing problems. Because cloning occurs naturally within bee populations, scientist intervention would not have to be as intrusive as it would with other species. As scientists continue to solve the problems the honeybee faces, cloning might prove to be just the right type of innovative thinking needed to correct the impact of human interference on nature.

Reference

Walsh, Bryan. 2013. "Beepocalypse Redux: Honeybees Are Still Dying—and We Still Don't Know Why." Time Magazine. http://science.time.com/2013/05/07/beepocalypse-reduxhoney-bees are-still-dying-and-we-still-dont-know-why/#ixzz2n4MaJtPL. Accessed on January 15, 2015.

Jeremy Summers is a freelance science writer whose work has appeared in Forbes, RealClearScience, *and* Truth about Trade, *among others. He lives and works in North Carolina and is a graduate of Appalachian State University.*

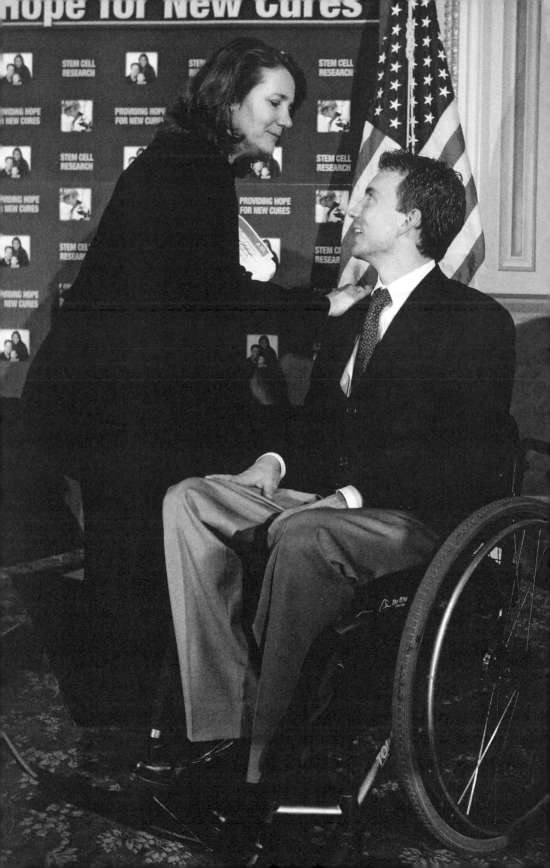

Introduction

In many regards, the history of cloning is a biographical tale. Learning about the developments of new ideas, new technologies, and new controversies is the story of the way people and organizations have lived their lives, carried out their research, dealt with their colleagues, and overcome challenges. This chapter focuses on a handful of the men and women who were key figures in the development of the science of cloning and its related field of stem cell research, as well as some of the important organizations that have and are contributing to these fields in a variety of ways.

Alliance for Regenerative Medicine

The Alliance for Regenerative Medicine (ARM) was founded in 2009 as a way of bringing together a variety of stakeholder groups in the area of regenerative medicine. Today, more than 140 member groups belong to the organization, representing business and industry, research institutions, foundations, academic institutions, investors, and patient advocacy groups. Some examples of the organizations that have become members of ARM include Blood Centers of America, Cell Therapies,

U.S. Representative Diana DeGette (D-CO) (L) talks with Jeff McCaffrey, who suffered a spinal cord injury in a car accident, during a news conference on stem cell research on Capitol Hill in Washington on July 19, 2006. (Jim Young/Reuters/Corbis)

Fisher BioSciences, GE Healthcare, Johnson & Johnson, Organogenesis, Pfizer, Vet-Stem, Toucan Capital, Novitas Capital, Abramson Cancer Center at Penn Medicine, the Cleveland Clinic, Texas Heart Institute, UC San Diego Stem Cell Program, University of Utah School of Medicine, American Association for Dental Research, Friends of Cancer Research, Missouri Cures, Parkinson's Action Network, and the Stop ALD Foundation. Individual memberships in the organization are not available.

ARM activities are carried out primarily through about a half dozen committees and working groups that include the following areas:

- Capital formation: Advocates for and recommends policies and practices in regenerative medicine to the federal and state governments.

- Communications and education: Works to enhance the visibility and credibility of regenerative medicine as a way of dealing with a wide variety of important health issues.

- Government relations and policy: Responsible for lobbying and otherwise interacting with the U.S. Congress, its members and staffs, and federal agencies with responsibility for healthcare issues

- Operations and governance: Responsible for all in-house governance and operational issues

- Regulatory: Interfaces with the U.S. Food and Drug Administration and other regulatory agency in the development of federal rules and regulations related to regenerative medicine

- Reimbursement: Deals with issues related to price structure and other payment issues involved in the research and development of regenerative medical products and their commercial sale

- Science and technology: A clearing house for scientific and technical information in the field of regenerative medicine, with special attention to the development and implementation of standards in the field of research and development.

ARM also maintains three technology sections to oversee and advise about research in the areas of tissue engineering and biomaterials, gene therapy and gene-modified cell therapy, and cell therapy. There is also a recently formed European Section that focuses on regenerative medicine issues that are specific to the European area.

A major focus of ARM activities is a number of conventions, conferences, and other types of meetings held throughout the United States and around the world. Examples of the types of programs sponsored and cosponsored by the organization are a CIRM (California Institute of Regenerative Medicine) webinar, the annual Biotech Showcase, the EU Advanced Therapies Investor Day, Medical Technologies Caucus Briefing, Global Control in Stem Cells program, and BioPharm America.

The ARM Web site is also a very useful source of information on a number of regenerative medicine-related topics. Its background page, for example, discusses the promise and potential of stem cell research and its applications in medicine, an industry snapshot section that describes the type of work that industry is conducting in the field of stem cell research, clinical testing information of various applications of stem cell therapies, clinical milestones in the development of regenerative medicine, information about acquisitions and financing in the regenerative medicine industry, and the economics of regenerative medicine.

Of particular interest is also an extended section on the applications of stem cell therapies for a number of specific diseases, including amyotrophic lateral sclerosis, Alzheimer's disease, autoimmune disorders, cardiovascular disease, diabetes, musculoskeletal disorders, ocular disease, Parkinson's disease, spinal cord injury, and stroke. The media center section of the Web page offers a variety of publications of interest both to members of the industry and the general public, including the Annual Industry Report and an application form for the organization's regular newsletter. The section also contains dozens of articles on the status of stem cell research and its medical applications.

American Anti-Vivisection Society

The American Anti-Vivisection Society (AAVS) is the oldest nonprofit organization in the United States devoted exclusively to the elimination of animal testing in scientific and biomedical research. The organization was formed in response to the growing use of animal-based research in the United Kingdom and the United States in the late 19th century. It took as its model the young National Anti-Vivisection Society founded in England in 1875. Although based on the English model, the roots of AAVS actually preceded the groundbreaking Cruelty to Animals Act of 1876 in Great Britain by a decade. In 1866, Philadelphia businessman Colonel M. Richards Mucklé announced his intention to establish a law enforcement society aimed at protecting animals from the worst abuses to which they were then exposed, a society that became the Philadelphia Society to Prevent Cruelty to Animals. Shortly thereafter, the wives of two members of that early group, Caroline Earle White and Mary Frances Lovell, decided to form a women's auxiliary, the Women's Branch of the PSPCA (WBPSPCA), which survives today as the Women's Humane Society of Pennsylvania (WHSP). As White and Lovell began to hear of the growing popularity of animal experimentation in the United States at the end of the century, they decided to form yet another group with the very specific objective of "preventing torture in the labs" of research scientists. That decision led to the creation of the American Anti-Vivisection Society in 1883.

AAVS was not slow in addressing its agenda among legislators. In 1885 it submitted its first bill before the U.S. Congress, a Bill to Restrict Vivisection. The bill was defeated, as were virtually all of the organization's legislative efforts over the next half century. The one success it points to during this period was blockage of an attempt to overturn the so-called 28-hour-rule, which provided for the humane care of animals in transit across state lines. On its Web site, AAVS admits that these early legislative efforts were "not always successful," but the organization was successful in other ways, primarily by making the general public aware of the moral and ethical problems posed by the growing use of animals in scientific and biomedical research. In fact, it was not until passage of the Animal Welfare Act in 1966 that the AAVS began to experience concrete legislative accomplishments in realization of its mission to reduce or eliminate the use of animals in research and testing.

Today, AAVS campaigns continue to focus on animal experimentation and animal testing, although they have expanded to include a number of other related topics. The End Animal Cloning project, for example, is an attempt to convince federal agencies to end and prohibit the cloning of animals for food production and other purposes, arguing that such practices not only pose a threat to public health but also represent a health risk to cloned animals themselves. Similarly, the Stop Animal Patents campaign is an effort to get governmental agencies from approving patents for the development of certain types of altered animal species, such as rabbits whose eyes have been intentionally damaged for product testing purposes. The association points out that more than 650 patents have already been granted for animals that have been altered in one way or another to make them more useful for research projects.

Most AAVS campaigns continue to focus on animal experimentation and animal testing issues. For example, its Ban Pound Seizure campaign aims to completely eliminate the practice of requiring or allowing public shelters to release unclaimed animals to researchers for experimentation. Only two states, Ohio and Oklahoma, still have laws that require shelters to release animals, and the AAVS admits that the problem is a "small, but troubling" part of the anti-vivisection challenge. Still, it has its goal the total elimination of such laws in all their forms and the complete protection of all animals who find their way into shelters. Similarly, the Animal Welfare Act campaign is an ongoing effort to get the U.S. Department of Agriculture (USDA) to include rats, mice, and birds in the list of species protected by the Animal Welfare Act of 1966. AAVS and other animal welfare groups have had some success in this campaign, but it has continually been counteracted by greater success of research groups, such as the National Association for Biomedical Research (NABR).

The AAVS Web site is a rich source of information on a variety of animal welfare topics, including basic information about the use of animals in research and the moral and ethical issues involved in such practices; alternatives to the use of animals in research; the role of animal experimentation and testing in educational programs and options available to students who do not wish to participate in such exercises; the use of animals in product testing; and laws relating to animal experimentation on a federal and state level. AAVS's primary publication for the general public is its quarterly magazine, *AV Magazine*, which can be downloaded from its Web site at http://www.aavs.org/site/c.bkLTKfOSLhK6E/b.8127027/k.921F/AV_Magazine.htm#.UC-_5d38u5J. The organization also publishes in print and electronic form a number of reports on topics such as "Animal Cloning: Animal Health Risk," a commentary on the Food and Drug Administration study on the risk to animals posed by cloning; "Primates by the Numbers," on the importation and use of nonhuman primates for research in the United States; "Dying to Learn," the report of a study on the use of dead and live animals by U.S. colleges and universities for classroom instruction; and "Genetic Engineering and Animal Welfare," which outlines the risks posed by genetic engineering for both human consumers and the animals produced by such research.

Americans for Cures Foundation

The Americans for Cures Foundation (ACF) was previously known as the Alliance for Stem Cell Research. The group was formed for the purpose of advocating and campaigning for California's Prop 71, which created the California Institute for Regenerative Medicine and authorized a $3 billion bond to support the work of that organization for the first 10 years of its existence. ACF reorganized as a federal 501(c)3 nonprofit formation to continue its efforts on behalf of what it claims to be the 100 million Americans who had diseases and disorders that can eventually be treated and often cured by the use of stem cell technologies.

ACF was founded by Robert N. Klein, a highly respected and successful financier and real estate developer. Klein became interested in healthcare issues when his youngest son, Jordan, was diagnosed with type 1 diabetes in 2001. At that point, Klein became active in advocating for diabetes research program and for patient issues, an effort for which he eventually received statewide and national recognition. For example, he received the Biotech Humanitarian Award in 2010 from the Biotechnology Industry Organization (BIO) for his efforts on diabetes and stem cell research. Klein is currently a director of the ACF.

On its Web site, ACF has listed a number of achievements of which it is most proud in its short history. These accomplishments include its role in the passage of Prop 71 and its later successful implementation as the California Institute for Regenerative Medicine (CIRM); its cohosting of a workshop on best practices in stem cell research in 2004; its hosting of four community-centered workshops in 2005 to inform the people of California about the planned work of the CIRM and to ask questions about the agency; its production of a DVD and film series on Spotlight on Disease, describing some of the ways in which stem cell research might develop treatments for a number of diseases; its work to implement a proposed international treaty banning all forms of human cloning in 2005; and the development of a national speaker network, designed to make experts on stem cell research available for presentations across the country.

The ACF Web site contains a number of useful articles, links, and other resources for those wishing to gain further information about stem cell research. The Press Room section, for example, provides resources categorized as opinion pieces, talking points, and press releases on stem cell research. A news section also includes more than 120 articles that have appeared in the press on various aspects of stem cell research, such as a listing of research rules on stem cell research, recent research on the use of stem cells for cancer treatment, a breakthrough in the use of stem cells in the treatment of diabetes, discoveries in methods for producing stem cells from alternative sources, and the use of stem cells for gene therapy research and treatment. The Web site also provides a well-written and very useful section on basic facts related to stem cell research arranged as articles on Stem Cell Facts, Facts about Reprogrammed Skin Cells (iPSCs), Frequently Asked Questions, and a stem cell glossary.

The Americans to Ban Cloning Coalition

The Americans to Ban Cloning Coalition (ABCC) was founded in 2002 to campaign against all forms of human cloning for whatever purpose. The organization provides very little information to the general public about its history, its organizational structure, its staff, or other administrative information. Perhaps the most widely mentioned name associated with ABCC is Richard M. Doerflinger, who has been involved in a number of abortion, stem cell, cloning, and related reproductive issues since 1980. He is currently associate director of the Secretariat of Pro-Life Activities of the United States Conference of Catholic Bishops, a post he has held for many years. Doerflinger is often the "face" of ABCC, writing articles, making speeches, testifying before governmental committees, and making other public presentations about cloning and related issues.

In its founding statement, ABCC called for "a global ban on human cloning for any purpose." The statement listed 10 reasons that it opposed this "gross misuse of cloning technology," including its purported "commodification and possible commercialization of human life"; creation of "a class of human beings who exist not as ends in themselves, but as the means to achieve the ends of others"; creation of "unnecessary dangers to the life and health of both child and mother"; and upsetting the social order by confounding the meaning of parenthood and confusing the identity and kinship relations of any cloned child, "The statement also argued that human beings have the right "not to be created as objects of experimentation," or to be a "gateway technology for the further genetic manipulation and control of human beings." In addition, the statement suggested that human cloning might "lead to the reproduction of living or deceased persons without their knowledge or involvement," and it would constitute "an affront to the inherent dignity and individuality of human life." Cloning, the statement concludes, also tends to mislead those who are "grieving the loss of a spouse, friend, or relative by promising what it cannot deliver, the bringing back of a deceased loved one."

The statement concludes with a very important point that explains why the organization is opposed to *all* forms of cloning, both therapeutic and reproductive. It claims that once *any* form of cloning is permitted, the opportunity for reproductive cloning will always be available. It argues that the term *therapeutic cloning* is misleading and prejudicial "since it has not been shown that cloning is necessary for or useful in the production of human therapies."

The primary value of the organization's Web site is the large collection of opinion pieces about cloning that have been produced by ABCC spokespersons and representatives over the years. These pieces are available on the Web site under the headings of Media, Resources, and News, although there is not a great deal of difference with regard to which pieces occur in which sections. Some examples of the materials available on the Web site include a number of press releases from the years 2002 to 2004 by groups such as the U.S. Conference of Catholic Bishops, the Roman Catholic Archdioceses of Newark, the

Traditional Values Coalition, the Holy See, and the National Right to Life Committee; testimony before the U.S. Congress and other legislative bodies, a collection of commentaries on cloning, and a cloning "fact of the day" archive from the organization's earlier history; and a rather limited collection of news articles taken from the Bioethics.com Web site. The articles listed under these categories are also classified separately under one of three major groups: science, law and policy, and commentary and analysis.

The ABCC Web site appears to be quite dated, with very little information from what seems to be its peak period of activity between 2002 and 2004. But the information available here is very useful because it provides a good view of the position of individuals and groups who are opposed to all forms of cloning, positions that tend to persist today among many individuals and groups.

Biotechnology Industry Organization

The Biotechnology Industry Organization (now known generally as BIO) was formed in 1993 as the result of the merger of two earlier biotechnology organizations, the Industrial Biotechnology Association and the Association of Biotechnology Companies. The merger was brought about largely through the efforts of attorney and businessman Carl B. Feldbaum, who then served as president of the new organization until his retirement in 2004. BIO claims to be the world's largest association of biotechnology companies, covering fields as diverse as industrial enzymes, genetically modified foods, pharmaceuticals and drugs, and biofuels. As of early 2015, it represented more than 1,000 companies in the United States and around the world employing more than 1.6 million Americans. A list of current BIO members is available at https://www.bio.org/articles/bio-members-web-site-links.

BIO has two major focuses in its work: advocacy on behalf of biotechnology with governmental agencies at all levels and among the general public, as well as assistance to biotechnology companies that aids in their development. To achieve these goals, the organization divides its work into four major areas: emerging companies, food and agriculture, health, and industrial and environmental. The objective of the emerging companies area is to assist young and smaller companies who may not yet have developed successful products by advocating for favorable tax and other regulatory policies and encouraging a generally more favorable economic, social, and political environment in which those companies have a better chance of success.

The food and agriculture area focuses on similarly working for a favorable governmental and public environment for the rapidly developing area of genetically modified foods and crops, presenting the industry position on topics such as the regulation and labeling of GMO foods and cops. The health component of BIO's work aims to improve laws and regulations and public opinion about the use of genetic engineering in a variety of health-related areas, such as healthcare policy and programs, climate issues, preparedness of pandemics and biowarfare, and personalized medicine. The industrial and environmental focus of BIO's work is aimed at educating the public and policy makers about the ways in which biotechnology can contribute to making human lives and the environment "cleaner, safer, and healthier."

Within these four broad areas, BIO has created a number of specialized areas of work that include, for health care: biopharmaceutical manufacturing and distribution, biosimilars, drug discovery and development, Food and Drug Administration reviews, personalized medicine, Prescription Drug User Fee Act, reimbursement and health policy, and vaccines and biodefense; for emerging companies: capital formation, funding, international market access, technology transfer, and therapeutic discovery project credit; for food and agriculture: animal biotechnology, plant biotechnology, and sustainable agriculture; and for industrial and environmental: advanced biofuels, algae, climate change, marine biotechnology, renewable chemicals and biobased materials, renewable specialty chemicals, and synthetic biology.

The BIO Web site is an extraordinarily complete and useful resource for virtually every issue related to biotechnology with pages on the U.S. Food and Drug Administration, Patents and Trademarks Office, Securities and Exchange Commission, Council of State Bioscience Associations, health care reform, the Sarbanes-Oxley Act, and the relationship of all these agencies and acts to the role of biotechnology in today's world. The site also provides useful information on basic topics in the field of biotechnology of interest and value to everyone from the neophyte to the professional in the field.

Among its many fields of specialized interest, BIO has taken a number of strong stands on the topic of cloning, in all case expressing its clear disapproval of human reproductive cloning, while pointing out the numerous benefits of therapeutic cloning and the role of biotechnology in achieving those benefits. See, as an example, the Web page on Frequently Asked Questions about Animal Cloning at https://www.bio.org/articles/frequently-asked-questions-about-animal-cloning and human reproductive cloning versus cloning of human cells at https://www.bio.org/media/press-release/human-reproductive-cloning-v-cloning-human-cells.

BIO sponsors a number of national, regional, and international conferences, the largest and most important of which is the annual BIO International Convention, held in 2015 in Philadelphia. That conference drew nearly 16,000 attendees, including about 2,500 CEOs, from all 50 states and 70 foreign countries. Other conferences sponsored by the organization include the BIO Asia International Conference, BIO Convention in China, BIO Latin America Conference, BIO Investor Forum, Livestock Biotech Summit, World Congress on Industrial Biotechnology, and Pacific Rim Summit on Industrial Biotechnology and Bioenergy.

For those without an extensive background in biotechnology, BIO provides a number of very useful Backgrounder publications in topics such as the BIO Statement of Ethical Principles, Genetically Engineered Animals: Frequently Asked Questions, Current Uses of Synthetic Biology, Synthetic Biology Explained, Glossary of Agricultural Biotechnology Terms, How Do Drugs and Biologics Differ?, and Background Information on Plant Biotechnology.

Brigitte Boisselier (1956–)

Boisselier was perhaps most famous in the early 2000s when she was science director of Clonaid, a scientific organization created by the Raëlian movement for the purpose of cloning a human child. On December 27, 2002, she appeared before a press conference in Florida to announce that the first cloned baby, named Eve, had been born at an undisclosed location. The announcement produced considerable furor in the scientific and general community and was later found to have been a fraudulent report promoted, leaders of the Raëlian movement said, to obtain publicity for their organization and their work in cloning.

Brigitte Boisselier was born, according to her Facebook page, in the small village of Champagne-Ardennes in the east of France on April 3, 1956. She claims to have been something of a child prodigy who taught herself to read at the age of four. She became interested in the study of chemistry early on in her life, and in particular in its application to the study of health and medical problems. She claims that she began writing letters to a variety of international organizations as early as the age of 14 in an attempt to obtain funding that would allow her to study in the United States. When those efforts failed, she enrolled at the University of Dijon in 1971, at the age of 15, where she majored in premedicine. She was eventually granted a master's degree in biochemistry at Dijon in 1975, followed by a PhD in physical chemistry also at Dijon. She then realized her dream of studying in the United States, matriculating at the University of Houston in 1982. She earned her second PhD there in electrochemistry three years later.

In 1985, Boisselier returned to France to join the industrial gas company, Air Liquide, where she served as vice director of research and development. She held that post for 13 years before leaving the company because of differences over her support of human cloning and religious beliefs. By that time, Boisselier, born a Catholic, had converted to the Raëlian religion, a belief system that holds that the human race on Earth was created by a group of alien scientists known as Elohim ("those who came from the sky") using procedures that we would now call genetic engineering. Obviously modern research on cloning and genetic manipulation is a crucial part of the Raëlian beliefs.

By 1992, Boisselier had become intrigued by the promise of human cloning and decided to convert from Roman Catholicism to Raëlianism. Before long, her new beliefs and participation in the new religion began to produce conflict with Air Liquide, and the company terminated her because her new work with Raëlianism was distracting her from her work at the company. She later sued the company for wrongful discharge and won her case, which included a settlement in the amount of $30,000.

After leaving Air Liquide, Boisselier devoted nearly all of her professional time to a new organization that Raël (the former French race driver Claude Vorilhon) had founded in 1997, Clonaid. Clonaid had been created for the specific purpose of using new cloning technology to produce the first cloned human baby. Boisselier was eventually appointed scientific director at Clonaid, as well as becoming an important member of the Order of Angels, the upper echelon of the religion's organizational structure. She was also named a bishop of the faith.

Boisselier's new commitment to Raëlianism and Clonaid created tensions in her own personal life. She and her husband separated, with the court assigning their youngest child to his father. Problems also arose between Boisselier and her parents over her participation in Raëlian activities, although her eldest daughter also converted to the new religion. In 2000, Boisselier and a second son moved to Canada just before accepting an appointment to teach chemistry at Hamilton College in Clinton, New York. Again, her work on behalf of Raëlianism and Clonaid caused conflicts with her new appointment, and she resigned from her post in 2001.

By that time, Boisselier claims that she was making substantial progress in bringing together a research team to carry out the cloning of a human baby that had become a passion for her and other Raëlian followers. Her December 2002 announcement represented a culmination, granted false one, of that effort. The failure of Boisselier and other Raëlians to provide evidence for the birth of Eve by no means meant the end of her human cloning experiments. By 2004, she claimed that Clonaid had already cloned 13 babies worldwide and that she and her colleagues were founding a new program, called Stemaid, that remains active today and claims to provide the technology needed to treat a range of diseases and conditions from aging and Alzheimer's disease to multiple sclerosis and stroke (see www.stemaid.com).

In 2006, Boisselier also founded another organization called Clitoraid, designed to provide medical, psychological, and other forms of assistance to women who had undergone genital mutilation. On her Facebook page, she points out that her faith in the advances available through cloning is illustrated by her having been one of the first human volunteer in a number of clinical trials of new stem cell research procedures.

Robert W. Briggs (1911–1983)

Briggs's place in the history of cloning and biology rests on a classic line of research that he conducted in 1952 with colleague Thomas J. King, in which the two men successfully transplanted nuclei from embryonic cells of living frogs into enucleated oocytes. They found that the nuclei selected from early embryos used in this research resulted in the growth and development of normal typical frogs of the same species, whereas those taken from more advanced embryos generally failed to grow and develop as new embryos. In addition to this now-famous experiment, Briggs pursued many other lines of research, perhaps most importantly on the role of maternal genes in the development of an embryo.

Robert William Briggs was born in Watertown, Massachusetts, on December 10, 1911. Both his mother and brother died of tuberculosis before Robert reached the age of two, so he was raised by his grandparents in Epping, New Hampshire, with whom he lived for the next 16 years of his life. He later told biographer Marie A. Di Barardino that he became interested in nature at an early age and made collections of plants, minnows, frogs, worms, and other living creatures. He never considered making a career of biology, however, because it never occurred to him "that one could earn one's living this way." His first job was during the summer at a local shoe factory when he was only 14 years old. He also earned money during the winters as a banjo player for a local band.

After graduating from high school, Briggs moved to Boston, where he enrolled as a business administration major at Boston University (BU), operating on the assumption that he would end up working in some type of business for the rest of his life. He discovered that he was not really very interested in business, however, and decided to enroll also in some science class in the College of Liberal Arts at BU. By the time he graduated in 1934 with a BS in biology, he had essentially committed himself to a career in science. He decided to continue his studies at Harvard University, from which he received his PhD in biology in 1938. He then accepted an appointment at McGill University in Montreal, where he began his research on embryogenesis, with special attention to the development of tumors in developing frogs.

In 1942, Briggs left McGill to accept an appointment as researcher at the Lankenau Hospital Research Institute in Philadelphia. It was at Lankenau that Briggs and King carried out their research on the transplantation of undifferentiated embryonic nuclei into an embryonic cell from which the nucleus had been removed, finally carrying out the "fantastical experiment" suggested by Hans Spemann in 1938 and paving the way for a long line of later somatic cell nuclear transfer (SCNT) experiments that have been central to the development of cloning technology.

In 1956, Briggs left Lankenau (which by that time had become the Institute for Cancer Research and was later to be renamed the Fox Chase Cancer Center) to become professor of zoology at Indiana University, where he remained until his retirement in 1982 with the title of professor emeritus of zoology. He lived only a year after his retirement, dying of kidney cancer in Indianapolis on March 4, 1983. Briggs was elected to the American Academy of Arts and Sciences in 1960 and to the National Academy of Sciences two years later. In 1973, he was awarded the Charles Leopold Mayer Prize of the French Academy of Sciences.

The Center for Bioethics and Human Dignity

The Center for Bioethics and Human Dignity (CBHD) was founded in 1993 by a group of Christian bioethicists who were concerned because of a lack of input from explicit Christian scholars with regard to a number of important issues in bioethics developing at the time. The group's first major action was calling a conference in May 1994 on "The Christian Stake in Bioethics," at which time CBHD was formally established. In 2007, the organization established a formal relationship with Trinity International University, an evangelical Christian institution located in Deerfield, Illinois. The center offers courses and continuing education programs in the field of bioethics, but does not offer formal certificates or degrees, which are instead offered by the university's master of arts in bioethics degree program and in related degree programs, such as its master of arts and master of divinity degree programs. CBHD is a 503(c)3 tax-exempt organization that is supported by charitable contributions, memberships, and income from its events.

CBHD's activities focus on about 20 specific topics, including bioethics in general, biotechnology, clinical and medical ethics, cloning, disability ethics, emerging technology, end of life, genetic ethics, global bioethics, healthcare, human dignity, human enhancement, neuroethics, organ donation and transplantation, public health, public policy, reproductive ethics, research ethics, stem cell research, and women's health. For each of these topics, the organization provides useful information on the topic in the form of a bibliography and a summary of most popular and most recent resources on the subject. (The available information on some topics is, as might be expected, more extensive on some topics than on others.)

The organization's Web page provides extensive background information explaining its strong opposition to all forms of cloning, both reproductive and therapeutic. The basic background for that position is provided in a 2001 position paper by John F. Kilner, president and CEO of CBHD, as well as Franklin Forman, chair of ethics, professor of bioethics and contemporary culture, and director of the Bioethics Program at Trinity International University. The Web page also provides links to a number of articles on cloning, such as "Bioethics at the Box Office: Cloning Edition," "Cloning Bibliography," "Cloning Annotated Bibliography," "The Trouble with Hwang: The Ethical and Scientific Problems of Therapeutic Cloning," "Embryonic, Fetal, and Post-Natal Animal-Human Mixtures: An Ethical Discussion," "Human Cloning," "Why Human Cloning Must Be Banned Now," and "Human Cloning: The Need for a Comprehensive Ban."

Probably the most important single document provided by the center is the last of those listed earlier, "Human Cloning: The Need for a Comprehensive Ban." That document lays out in great detail the five basic reasons that all types of cloning experiments should be prohibited, for both reproductive and therapeutic purposes. The article concludes with the warning that "[f]ailure to adopt such a ban will result in scientific, ethical, and legal failures—the scope and consequences of which will be of great magnitude." In support of its argument, the article also provides 46 endnotes as references for its position.

The center sponsors, cosponsors, and contributes to a number of events in support of its efforts in opposition to cloning. Examples of such events from 2014 include its annual summer conference on bioethics, "Bioethics in Transition," held at Trinity International University, and webinars on topics such as "Healthcare Ethics," "Evidence-Based Best Practices in Reducing Maternal Mortality," "A Sustainable Model for Reducing Maternal Mortality: Evidence from the SHARE Project in Sub-Himalayan India," and "Father's Vital Contribution to Improving the Health & Safety of Mothers and Newborns: Transforming Communities in Nicaragua through the Education of Fathers."

The center's primary regular production is its quarterly publication, *Dignitas*, which describes itself as "a vehicle for the scholarly discussion of bioethical issues from a Judeo-Christian Hippocratic worldview, updates in the fields of bioethics, medicine, and technology, and information regarding the Center's ongoing activities." The journal is available to members of the center.

Center for Food Safety

The Center for Food Safety (CFS) was founded in 1997 by attorney, environmental activist, and consumer advocate Andrew Kimbrell, who still serves as executive director of the organization. CFS was founded as a nonprofit organization for the purpose of working to protect human health and the natural environment by opposing the use of harmful food production technologies and by promoting organic and other forms of sustainable agriculture.

The center organizes its activities around 10 major themes: genetic engineering, seeds, pollinators and pesticides, animal cloning, food and climate, aquaculture, food safety, factory farms, organic and beyond, and nanotechnology. For each of these topics, the CFS Web site provides detailed information about the issue and suggestions for ways in which consumers can become involved in actions to resolve specific issues. For each topics there is also a list of publications and resources that include reports on the topic, legal actions taken with respect to the issue, policy statements by various organizations, testimony provided by CFS representatives and other individuals, and news about recent development with regard to the issue. The Web page on animal cloning, for example, provides separate Web pages called "About Cloned Animals" (which provides basic information on the topic), "Human Health" (which reviews the scientific evidence—or lack of it—on the health effects of eating cloned animals), "Animal Welfare" (which discusses the impact of cloning on animals that have been produced by the procedure), "Biodiversity" (which raises questions about the ways in which cloning and cloned animals may have deleterious effects on environmental biodiversity), and "Government Regulations" (which reviews laws and administrative rules that may have impact on the production, sale, and consumption of cloned animals).

The center has also developed a number of initiatives to which members and interested readers are invited to participate. Examples of those initiatives include campaigns to get Lowe's and Home Depot to stop selling pesticides that are toxic to honeybees, to stop the development of genetically engineered fish, and to stop GE crop field tests and petitions to get the Environmental Protection Agency more involved in efforts to protect honeybees, to urge the retailer Trader Joe's to stop selling meat containing antibiotics, and to convince the U.S. Department of Agriculture to ban the growing and sale of genetically engineered apples.

CFS also makes available to members and interested others a wide range of print and electronic publications in the form of fact sheets, newsletters, reports, policy statements, testimonials, and legal actions. Examples of fact sheet topics that are available on cloning are "Compilation and Analysis of Public Opinion Polls on Animal Cloning" and "Cloned Food: Coming to a Supermarket Near You?" The center has also produced a report entitled "Not Ready for Prime Time: FDA's Flawed Approach to Assessing the Safety of Food from Animal Clones." The organization's primary newsletter is *Food Safety Now*. It also provides news electronically to subscribers on its True Food Network.

A final section of the Animal Cloning Web page provides copies of documents prepared by CFS for legislative committees, administrative bodies, and other law-and policy-making bodies at the federal level. Examples are written comments on a 2007 FDA proposal to allow milk and meat products from cloned animals into the U.S. food supply unlabeled, a 2007 action by the National Organic Standards Board opposing the inclusion of cloned animals or their offspring under the existing organic standards, and the 2003 study concerning the risk of eating food from cloned animals, as well as copies of testimony presented by representatives of the center at Congressional hearings on the use of cattle pedigrees to track and exclude cloned animals from organic production (2011) and labeling of food containing a product from cloned animals.

Compassion in World Farming

Compassion in World Farming (CIWF) was founded in 1967 by Peter and Anna Roberts, farmers in Hampshire who, based on their own experiences, became concerned about the growth of a new form of agriculture in which the well-being of farm animals and the environment became of increasing disinterest to large commercial farmers and commercial agricultural organizations. They began by calling a few friends and neighbors to a meeting held at their kitchen table to draw up plans as to how they could work to resist the most undesirable aspects of factory farming. The organization has since grown to become an important international spokesperson for the humane care of animals in modern farming operations. In 2011, CIWF opened a branch office in the United States. The organization also has branches specifically in France, The Netherlands, Italy, and Poland. In addition to its general membership CIWF also recognizes special groups of members who are "high-profile supporters"; "celebrity ambassadors"; and "visionaries"; and notable individuals from the arts and media, civil society, faith and ethics, sports, business, and academia.

CIWF uses a variety of programs and initiatives to promote their concern for farm animals. In 2007, it launched its food business team, which works with food producers, manufacturers, and retailers to include animal welfare as part of its operating procedures. CIWF has since created a number of awards for companies that make progress in this area. The first such award was the Good Egg award, given annually since 2007 to companies who produce their eggs from cage-free chickens. The organization has since added also a Good Chicken Award in 2010, Good Dairy Award in 2011, and Good Pig Award in 2012. CIWF claims that 287 million farm animals will benefit each year from the adoption of more humane farming conditions promoted by the organization and adopted by the industry.

In addition to its educational and advocacy activities, CIWF has conducted and sponsored a very wide array of research studies on various aspects of the way farm animals are being treated in modern agricultural and dairy facilities. Some of the reports produced as a result of this research are the "Calf Forum Report"; "TTIP (Transatlantic Trip and Investment Partnership): A Recipe for Disaster"; "A Sustainable Food Policy for Europe"; "Fats, Farming, and Food"; "Antibiotics Briefing—Antimicrobial Resistance"; "Down to Earth Charter for a Caring Food Policy"; and "The Life of Dairy Cows." All of these, and nearly 100 other reports, are available on the organization's Web site for downloading and viewing. The Web site also makes available a collection of news articles related to compassionate farming issues intended for the general public and dealing with topics such as the California state ban on foie gras, the national campaign in the United States to reduce meat consumption, and the relationship between factory farming and animal cruelty.

In addition to its work with the agricultural and dairy industries, CIWP carries on an intensive educational program designed for both students and the general public. A key element of this program is the provision of a number of free resources from the organization's Web site. These resources include lecture notes (e.g., "Good Agricultural Practices—Pigs"), activities and work cards for young children (e.g., on the needs of farm animals), PowerPoint presentations (e.g., "Good Agricultural Practices—Pigs"), and discussion cards and worksheets (e.g., "How Should We Treat Farm Animals?").

The U.S. chapter of CIWF has adopted the issue of cloning as a special area of concern and produced a 2012 142-page report on the topic that is available at the organization's Web site for free downloading. The UK site also has an extensive selection of articles dealing specifically with the topic of cloning. Those articles are most easily accessed by using the Web site's search function for the term *cloning*.

Council for Responsible Genetics

The Council for Responsible Genetics (CRG) was formed in 1983 by a coalition of scientists, public and occupational health activists, and reproductive rights advocates in response to the increasing number and variety of applications of genetic engineering in everyday life. The founders' view was that both the scientific community and the general public were growing excited by the potential of genetic engineering to change agriculture, biological warfare, human and other animal reproduction, and a host of other aspects of human society, but that there was no place for people to go to learn about the social, economic, psychological, ethical, and other issues associated with these changes. They saw CRG as the first resource for such a balanced view of the impact of genetic engineering on human life.

One of the first areas in which CRG became involved was the potential use of genetic engineering in the development of biological weapons. In 1985, the organization organized a briefing for the U.S. Congress on the issue, and a year later it sponsored a special session at the annual meeting of the American Association for the Advancement of Science (AAAS) on the topic. CRG was also active in the writing and passage of the Genetic Information Nondiscrimination Act of 2008. Other accomplishments include publication of the book *Genetically Engineered Food: Changing the Nature of Nature* (Martin Teitel and Kimberely A. Wilson, Park Street Press, 1999); creation of a "Citizen's Guide to Genetically Modified Food"; promotion of the Safe Seed Campaign, a program through which seed manufacturers and distributors pledge to use only to nongenetically modified seeds; adoption and publication of a "Genetic Bill of Rights," a document designed to encourage international discussion about the types of human rights that have become endangered by the development of genetic engineering (available online at http://www.councilforresponsiblegenetics.org/projects/CurrentProject.aspx?projectId=5); publication of the book *Rights and Liberties in the Biotech Age: Why We Need a Genetic Bill of Rights* (Sheldon Krimsky and Peter Shorett, Rowman & Littlefield, 2005), which provides commentaries on the Genetic Bill of Rights; and sponsorship of a conference on Race and Genetics in June 2008 to discuss the racial and civil liberties implications of the development of DNA databases by governmental and legal agencies.

Among the issues with which CRG is currently dealing is the topic of cloning and human gene manipulation. The organization's Web site provides a wealth of background and advocacy information on the topic, including links to a number of media presentations by CRG staff and experts, an official position statement about human cloning by the organizations, background papers on the science and ethics of human cloning, and recommended articles on the topic.

Perhaps the organization's most valuable resource is its magazine *GeneWatch*, published about four times a year (sometimes more frequently, sometimes less frequently). Each issue of the magazine focuses on some specific aspect of genetic technology, such as cloning, race and genetics, gene patents, biolab safety, forensic DNA, bioengineering animals, or privacy in the age of genomics.

Diana DeGette (1957–)

DeGette has sponsored a number of bills calling for federal funding of research on stem cells. Her bills passed the Congress in 2005 and 2007, but failed to reach the floor of the House in 2009, 2011, and 2013. In announcing her introduction of the 2013 Stem Cell Research Advancement Act (with Representative Charlie Dent R-PA), DeGette pointed out that "We have the opportunity to treat, if not outright cure, some of the most horrendous diseases that people suffer from because of this research. . . . The Stem Cell Research Advancement Act puts in place the necessary framework needed to ensure this vital research is free from future political interference." DeGette has consistently been a strong advocate for a number of other social causes, including access to abortion and contraception, gun control, the 2009 Affordable Care Act, same-sex marriage, environmental protection, unemployment insurance, and veterans' healthcare.

Diana Louise DeGette was born on July 29, 1957, at Tachikawa Air Force Base in Tokyo, Japan, to Richard Louis DeGette and Patricia Anne Rose DeGette. Her father, later an architect and planner, was serving in the U.S. Air Force at the time. After the family's return to the United States, DeGette attended South High School, in Denver, Colorado, from which she graduated in 1975. DeGette then enrolled at Colorado College, where she majored in political science, earning her BA in 1979. She then continued her studies at New York University, from which she received her JD in 1982. From 1982 to 1984, DeGette served as deputy state public defender for the state of Colorado before going into private practice at the end of that period. In 1985 she married attorney Lino Lipinsky de Orlov, with whom she later had two children, Raphaela and Francesca.

DeGette's first foray into politics occurred in 1992 when she ran as a Democrat for the Colorado House of Representatives from southeast Denver. She won that and the ensuing election. She was also elected assistant minority leader during her first term in the house. DeGette then decided to run for the U.S. House of Representatives in 1995 when long-time incumbent and liberal icon Patricia Schroeder announced her retirement as representative from Colorado's First District. DeGette won that race against Republican Joe Rogers by a margin of 57 to 40 percent. She has since been reelected 11 times, always with majorities in the 60- and 70-percent range. DeGette has served as a chief deputy whip for the Democratic Party since 2005.

Jay Dickey (1939–)

Dickey represented the Fourth Congressional District of Arkansas from 1993 to 1999. He was elected and served as a strong conservative in a district that had never previously elected a Democrat to the U.S. House of Representatives. While serving in Congress, he espoused and fought for many traditional conservative causes, such as gun rights, antigay rights, and anti–stem cell research programs. In 1995, Dickey coauthored with Senator Roger Wicker (R-MS) an amendment to the appropriations bill for the Department of Health and Human Services, prohibiting that department from providing federal funding for any form of stem cell research. The amendment was later adopted every year as an amendment to that appropriations bill until it was found to be unconstitutional by the

U.S. Court of Appeals for the District of Columbia Circuit in 2011.

Jay Woodson Dickey Jr. was born in Pine Bluff, Arkansas, on December 14, 1939, to Jay Woodson Dickey Sr. an attorney, and his wife Margaret. He attended Pine Bluff High School, from which he graduated in 1957. He then enrolled at Hendix College in Conway, Arkansas, which he left after one year, transferring to the University of Arkansas in Fayetteville (UA-F). He received his BS from Arkansas in 1961 and then continued his studies at UA-F law school, where he received his JD degree in 1963. After graduation, he opened a private practice in Pine Bluff, which he maintained until 1968, when he became city attorney for the city. After two years in that position, he returned to private practice in Pine Bluff. In addition to his law career, Dickey was a successful businessman, operating a number of Taco Bell and Baskins and Robbins locations in Pine Bluff.

In 1992, Dickey decided to try his hand at politics, entering the race for representative in Arkansas's Fourth Congressional District. He might have appeared to have little chance in the race, because the district had never elected a Republican to the seat. Dickey's opponent in the election had been accused of scandal in office, however, and Dickey won the election by a vote of 52 to 48 percent. He was later reelected in 1994, 1996, and 1998 by comfortable margins. His 2000 campaign against Democrat Mike Ross was unsuccessful, however, and, in a return match two years later, he was handily defeated by Ross by a vote of 61 to 39 percent.

After leaving office, Dickey opened his own lobbying firm in Washington, D.C., which survived until 2009. During its period of operation, the firm represented companies and agencies such as the Arkansas Hospitality Association; the Jefferson Regional Medical Center; the city of Monticello, Arkansas; the Ludington Area Chamber of Commerce; and Pine Bluff Sand & Gravel. In its most successful year (2004), the company had lobbying income of $565,000.

In recent years, Dickey has eschewed some of the causes for which he fought most strongly while in the Congress. For example, he and Mark Rosenberg, president and CEO of the Task Force for Global Health, wrote an op-ed for the Washington Post newspaper in which they called for stronger federal action against gun violence. The article represented, according to the Encyclopedia of Arkansas History and Culture, "a 180-degree political turn by calling for the restoration of public funds for research into gun safety." (Dickey's change of heart was supposedly prompted at least in part by the mass shooting at Aurora, Colorado, in July 2012.)

Hans Driesch (1867–1941)

Driesch is sometimes credited as having conducted the first experiments on the cloning of animals, research carried out in the early 1890s. In this research, Driesch separated the two cells of an early sea urchin embryo simply by shaking them and then observing that each cell eventually developed normally into a new and complete sea urchin embryo. Some historians suggest that this research, while important, is not really an example of cloning (since only two clones were produced), but of the more limited process of twinning, in which only two clones are produced. Driesch's work is, nonetheless, generally regarded as an important early step in the history of animal cloning.

Hans Adolf Eduard Driesch was born in Bad Kreuznach, Prussia, on October 28, 1867. His father, Paul Driesch, was a successful gold merchant in Hamburg, and his mother, Josefine Raudenkolb Driesch, an enthusiast of natural history who maintained a substantial "zoological garden" in the family home that consisted not only of a wide variety of plants but also of specimens of many kinds of animals, including monkeys, parrots, lizards, snakes, and alligators. Young Hans grew up in this setting and soon developed an avid interest in nature himself, an interest that was probably responsible for his later decision to devote his life to a study of the origins and development of living organisms.

Driesch received his secondary education at the Gymnasium Gelehrtenschule des Johanneums in Hamburg, after which he continued his studies at the University of Freiburg under the famous German embryologist August Weismann. After two semesters there, he transferred to the University of Jena, where he had the opportunity to study with the equally famous and highly admired (by Driesch) Ernst Haeckel and from which he received his doctorate in embryology in 1889. (Driesch had also taken a semester off in 1888 to study physics and chemistry at the University of Munich because he thought himself deficient in those two subjects, which were not at the time offered at Jena.)

After completing his studies at Jena, Driesch took a position as researcher at the Naples (Italy) Zoological Station, where he was to remain for the next decade. It was at Naples that he conducted his first experiments on the cloning of sea urchins and, in fact, conducted virtually all of the experimental research that he was to complete during his life. During the same period, however, he became increasingly interested in a different aspect of embryology, namely the theoretical and philosophical basis for understanding the process by which a single fertilized egg develops into a complete organism. He had begun this intellectual journey with an essentially mechanistic outlook on the question, a position held by Haeckel and his followers. According to that view, living organisms were essentially and for all practical purposes the same as nonliving matter, objects that could eventually be completely understood through the principles of chemistry and physics. Driesch gradually became more interested in a vitalistic interpretation of life, one that was based on the belief that living organisms possess some "vital force" that can never be understood by physical principles alone.

Although he had maintained a formal connection with Naples for an extended period of time, he was at the time independently wealthy and able to travel extensively to study and carry on his research. In fact, he was one of the rare scholars of the time who was rarely continuously associated with one university or another. A turning point in his life occurred, then, in 1907 when he was invited to give the Gifford lecture in natural theology at the University of Aberdeen in Scotland. Following the success of that lecture, and another in the same series a year later, Driesch was able to obtain an appointment as professor of natural philosophy at the University of Heidelberg, after which he held similar positions at the universities of Cologne (1919) and Leipzig (1921). In addition to his regular teaching posts, he was also visiting professor in a number of universities around the world, including China, South America, and the United States. In 1933, Driesch was forced to retire from his university teaching post by the new Nazi government, the first non-Jew to be so treated not because of his religion, but because of his opposition to the new National Socialist government. Driesch continued to write and speak about his research and his philosophical views, however, until his death in Leipzig on April 17, 1941.

Empresa Brasileira de Pesquisa Agropecuária (Embrapa)

Empresa Brasileira de Pesquisa Agropecuária (Brazilian Corporation of Agricultural Research; Embrapa) is a division of the Brazilian Ministry of Agriculture, Livestock, and Food Supply with the responsibility for carrying out original research on plants and animals with important roles in the Brazilian economy. The agency was established in 1973 in response to significant changes in government policies with regard to the development and use of national lands for the growth of crops and the development of the dairy industry that would provide the food necessary for the growing population. (For details on the history of the agency, see Geraldo P. Martha Jr., Elisio Contini, and Eliseu Alves, "Embrapa: Its Origins and Changes," http://ainfo.cnptia.embrapa.br/digital/bitstream/

item/81263/1/Embrapa-its-origins.pdf, accessed on January 25, 2015.) Embrapa has three major central offices in Brasilia; Manaus, Amazonia; and Campina Grande, Paraiba, along with about five dozen regional offices devoted to a variety of research and development activities, such as agriculture informatics, agrobiology, agroenergy, beef cattle, cassava and tropical fruits, coastal tablelands, coffee, cotton, dairy cattle, fishers and aquaculture, goats and sheep, grapes and wine, plant quarantine, satellite monitoring, soils, southern livestock, territorial management, and wheat.

Embrapa conducts well over a thousand different research projects at any one time in the general area of ecosystems, regions, portfolios, macroprograms, and units. These general areas are further subdivided into topics such as specific ecosystem (e.g., Amazonic, Cerrados, Coastal Areas, Extreme South, and Panatanl) or regions (Central-West, North, Northeast, South, and Southeast); animal health projects; aquaculture projects; palm oil research; native forest resources; nitrogen fixation; competitiveness and sector sustainability; institutional development; technology transfer and business communication; and specific crop programs, such as cattle, cotton, and maize and sorghum.

In 2012, Embrapa announced a joint program with the Brasilia Zoological Garden to begin cloning threatened and endangered species in Brazil. The agency has already produced its first cloned animal, a cow called Vitória, in 2001, and had begun to collect more than 400 tissue samples from dead animals. Some of the animals scheduled for possible cloning were the maned wolf (*Chrysocyon brachyurus*), black lion tamarin *(Leontopithecus chrysopygus)*, bush dog (*Speothos venaticus*), coati (*Nasua nasua*), collared anteater (*Tamandua tetradactyla*), gray brocket deer (*Mazama gouazoupira*) and bison (*Bison*). None of these animals is as yet classified as endangered, but the Embrapa/Zoological Garden program is hoping to get a head start on possible future declines of the species. The program's plan is to begin cloning animals and then keep them in captive facilities to replenish the native stock if and when that becomes necessary.

The Embrapa Web site is an excellent source of information about the agency itself as well as the natural history of Brazil. Information is available in both Portuguese and English, although some technical documents have not been translated from the original Portuguese. The Web site provides detailed news about the agency's activities, copies of reports of research conducted by staff, videos and images of the agency's works and the organisms with which it works, and background scientific information on a host of relevant topics.

Martin Evans (1941–)

Evans was awarded a share of the 2007 Nobel Prize in Physiology or Medicine for his research on embryonic stem cells in mice, carried out in association with Scottish development biologist Matthew Kaufman. The discovery of murine (mouse) stem cells was announced at almost the same time as the Evans-Kaufman research by Gail R. Martin, who was then affiliated with the University of California at San Francisco. After producing his breakthrough research on murine stem cells, Evans has spent most of his subsequent research on additional studies of the production, isolation, and use of mouse stem cells in modifying organisms for studies of embryogenesis and related phenomena.

Martin John Evans was born on January 1, 1941, in Stroud, Gloucestershire, England. In an interview with Desert Island Discs, Evans explained that his mother was a teacher, and his father the owner of a mechanical workshop. Evans also told the interviewer that he developed an interest in science as a young child, an interest that his parents encouraged. His early schooling took place at St. Dunstan's College (a middle school) in London, after which he matriculated at the University of Cambridge, which he attended with the benefit of a scholarship at Christ's College. He majored there in botany, chemistry, and biochemistry, earning his bachelor's degree in 1963. After graduation, he accepted a position as research assistant at University College London (UCL), from which he received his PhD in anatomy and embryology in 1969.

Evans stayed on at UCL after earning his degree, working in the department of genetics. He remained there for more than a decade, before finally realizing that he had probably gone as far as possible in the university's academic system. He decided, therefore, to look for other opportunities and in 1978 applied for a position with the department of genetics at the University of Cambridge. Two years later, he began a collaboration with Kaufman that was to result in their breakthrough discoveries on the isolation of murine stem cells. Kaufman left Cambridge in 1985 to assume the chair of anatomy at the University of Edinburgh in 1985, but Evans stayed on at Cambridge for another 14 years. Perhaps his most significant accomplishment during that period of time, beyond his normal schedule of research, was the cofounding of the Wellcome/CRC Institute for Cancer and Developmental Biology, since renamed the Gurdon Institute.

In 1999, Evans experienced what he described in his Nobel autobiography a "huge change of role" when he accepted an appointment as professor of mammalian genetics and director of the newly created School of Biosciences at the University of Cardiff in Wales, where he remained until his retirement from active research in 2007. (In 2013, the school renamed the building in which it is located as the Sir Martin Evans Building.) Two years later, he was appointed president of the University of Cardiff, and, in 2012, he was promoted to chancellor of the university.

In addition to his Nobel Prize, Evans has received a number of other honors and awards, including the March of Dimes Award in Developmental Biology (with Richard Gardner) (1999), Albert Lasker Award for Basic Medical Research (with Mario Capecchi and Oliver Smithies; 2001), Gold Medal of the Royal Society of Medicine (2009), Copley Medal of the

Royal Society (2009), and UCL Prize Lecture in Clinical Science (2011). He was made a fellow of the Royal Society in 1993 and knighted by Queen Elizabeth II in 2004. He has been given honorary doctorates by the Mount Sinai School of Medicine, the University of Bath, and University College London.

John D. Gearhart

Gearhart, working at Johns Hopkins University in Baltimore, followed James Thomson at the University of Wisconsin by less than a week in 1998 in announcing a significant breakthrough in the production, isolation, and culturing of human embryonic stem cells. Although the Wisconsin and Gearhart teams used similar research protocols, they began with different materials. While Thomson had used fertilized eggs produced by in vitro fertilization procedures, Gearhart's team obtained their stem cells from the immature gonads in fetuses that had been aborted. The important feature of this distinction is that Thomson's research did not that fall with the criteria that would have allowed federal funding (he depended on private funding for his research), while Gearhart's did.

John Gearhart was born in Western Pennsylvania, where he lived until he reached the age of six years. At that point, his father died, and while his mother and brother remained on the family farm, young John was sent to Girard College, a school for male orphans in Philadelphia. Gearhart remained at Girard for the next 10 years, completing his high school education there in 1960. He then enrolled at Pennsylvania State University where he planned to major in some field of agriculture or biological sciences. This interest grew out of his early days on a farm in Pennsylvania, but gradually became stronger as he pursued his study of the sciences. He once told an interviewer for the Academy of Achievement that his original plans upon entering Penn State were "the best pomologist in the world." (Pomology is the study of the cultivation of fruits, such as apples, pears, and oranges.) As time went on, however, he became more interested in the field of genetics, and, after receiving his BSc in biology from Penn State in 1964, he decided to continue his studies in genetics at the University of New Hampshire, from which he received his master's degree in genetics in 1966. He then moved on to Cornell University, where he was awarded his PhD in genetics and development in 1970.

After completing his degree at Cornell, Gearhart returned to Philadelphia, where he spent two years as a postdoctoral fellow at the Institute for Cancer Research in the laboratory of the eminent researchers Beatrice Mintz. Mintz is famous for having produced the first mouse chimera in the 1960s. After completing his postdoctoral research, Gearhart decided to stay on at the institute for three more years as a research associate. Then, in 1975, he accepted an appointment as assistant professor of anatomy at the University of Maryland School of Medicine in Baltimore, where he remained until 1980. He then moved to the Johns Hopkins University School of Medicine, also in Baltimore, with whom he has been affiliated ever since. He has worked his way up the academic ladder over the past three-plus decades, serving as associate professor and professor of cell biology, anatomy, gynecology, obstetrics, population dynamics, physiology, comparative medicine, biochemistry, and molecular biology at one time or another. He is currently adjunct professor at the Johns Hopkins University School of Medicine, James W. Effron University professor at the University of Pennsylvania School of Medicine, professor of animal biology at the University of Pennsylvania School of Veterinary Medicine, and professor of cell and developmental biology in obstetrics and gynecology at the University of Pennsylvania School of Medicine. In addition to his academic appointments, Gearhart has held and continues to hold a number of administrative and clinical appointments such as director of the preimplantation genetic diagnosis program (1997–2006), cofounder of the Institute for Cell Engineering (2002–2008), and cofounder of the Stem Cell Policy and

Ethics Program (2002–2008), all at Johns Hopkins, and director of the Institute for Regenerative Medicine at the University of Pennsylvania (2008–present).

In addition to more than 100 refereed publications, Gearhart is a widely popular public speaker on the subject of stem cell research and regenerative medicine. His more than five dozen invited lectures have taken him to locations as widespread as Monterey, California; Bethesda, Maryland; London; Laramie, Wyoming; Shanghai; Omaha, Nebraska; and Beijing.

The Genetics Policy Institute

The Genetics Policy Institute (GPI) was founded in 2003 by attorney Bernard Siegel, who was then probably best known for having filed suit in the Broward County Court to obtain a guardian for the purported firstborn cloned human child, Eve, as announced by the Clonaid company. (That claim was later to be judged to be fraudulent.) GPI's mission was and is "promoting and defending stem cell research and its application in medicine to develop therapeutics and cures for many otherwise intractable diseases and disorders."

The institute's primary activity is its sponsorship of the World Stem Cell Summit, first held in 2004 held at the University of California at Berkeley. In the same year, GPI sponsored an international meeting on human cloning at the United Nations headquarters in New York City. Later in 2004, the institute also held a Stem Cell Awareness Day at the Miller School of Medicine at the University of Miami and a Patients Press Conference at the United Nations headquarter in New York. GPI has continued to sponsor the annual World Stem Cell Summit ever since, along with a number of other RegMed (for regenerative medicine) Capital Conference that aims to bring together private investors with researchers and industry leaders in the field of regenerative medicine. GPI also presents a set of awards every year for individuals who have made significant contribution in the field of stem cell research and regenerative medicine. The awards are given for leadership, international leadership, education, advocacy, and inspiration.

Much of the institute's work is carried on through a variety of so-called initiatives. An example is Stem Cell Action, a group of nearly 100 organizations that lobbies the U.S. Congress and other policymakers to make stem cell research and regenerative medicine major health priorities. Members of the group include such diverse organizations as the A. Alfred Taubman Medical Research Institute at the University of Michigan, Alliance for Aging Research, American Academy of Neurology, B'Nail B'rith International, The Cell Transplant Society, Center for Inquiry, Cystic Fibrosis Research, Inc., Federación Argentina de Enfermedades Poco Frecuentes, London Regenerative Medicine Network, Texans for Stem Cell Research, and Wisconsin Stem Cell Now.

GPI also sponsors a variety of policy, legal, and education initiatives. Examples of the activities included in this area are the submission of briefs to appropriate courts in support of federal funding for stem cell research; successful lobbying of the United Nations to prevent adoption of a treaty calling for the prohibition of therapeutic cloning; lobbying of the National Institutes of Health on the development of its Draft Guidelines for Human Stem Cell Research; production of a variety of educational materials such as press releases, media interviews, and op-ed pieces in support of stem cell research; and development of coalitions with nearly two dozen like-minded groups, such as the Coalition for the Advancement of Medical Research, Americans for Cures Foundation, and Alliance for Regenerative Medicine.

In addition to its major annual production, the "World Stem Cell Report," an account of the annual convention, GPI produces four different newsletters, "360 Weekly Newsletter," dealing with business, policy, research, and advocacy issues; "Genetic Policy Institute Updates," focusing on the structure and activities of GPI, "World Stem Cell Summit Updates," providing information specifically on the annual conference itself; and "Stem Cell Action Updates," whose primary interest is in developments in the worldwide Pro-Cures Movement.

The GPI Web site also has an extensive collection of current and archival news items dealing with all aspects of the fields of stem cell research and regenerative medicine. Siegel and other staff members of GPI are also very active in participating in many different kinds of meetings relating to stem cell research and regenerative medicine. Examples of their schedule for 2015, for example, included participation at the Regenerative Medicine State of the Industry Briefing and Biotech Showcase in San Francisco, BIO International Convention in Philadelphia, Annual Meeting of the International Society for Stem Cell Research in Stockholm, and the Regenerative Medicine Essentials Course on The Fundamentals of the Future at Winston Salem, North Carolina.

John Gurdon (1933–)

Gurdon was awarded a share of the 2012 Nobel Prize in Physiology of Medicine for his research on somatic cell nuclear transplantation (SCNT). In 1958, Gurdon became the first person to successfully clone a frog by inserting the nucleus of an intestinal cell from a tadpole of the species *Xenopus laevix* into an enucleated egg of an adult frog, after which the egg developed normally, resulting in the birth of a complete normal frog. The experiment was important because it demonstrated that an adult cell retained all the genetic information required for the development of a complete and normal adult organism, a concept whose proof of which researchers had been pursuing for many decades.

John Bertrand Gurdon was born on October 2, 1933, in Dippenhall, Hampshire, England. His parents were William Nathaniel Gurdon, who worked as banker in India before taking early retirement while still in his 40s, and Elsie Marjorie Byass Gurdon, who had worked as a physical education teacher at a private American school in England. Gurdon enrolled at

Eton College (a secondary school) at the age of 13, but was just considerably to be a spectacular failure there. In what was probably the most famous document in his early life, his headmaster at Eton at one point wrote that "I believe Gurdon has ideas about becoming a scientist. On present showing, this is quite ridiculous. If he can't learn simple biological facts he would have no chance of doing the work of a specialist, and it would be a sheer waste of time both on his part and of those who would have to teach him." At the time, Gurdon ranked last among 250 students in biology, as well as at or near the bottom of every other class in which he was enrolled. Gurdon is said to have framed that report and hung it over this desk at work.

After graduating from Eton in 1952, Gurdon sat for entrance examinations in classics at Christ Church College, Oxford University. He was admitted to Christ Church, but only on condition that he enroll, ironically enough, in the department of zoology rather than pursing a course in Latin and Greek. As it later turned out, the college had already filled the available seats in classics, but had a number of empty spaces left in the sciences. As a result, Gurdon ended up in precisely the subject in which he had long been interested, but from which he had been discouraged in pursuing.

Gurdon remained at Oxford for the next eight years, earning both his bachelor's and doctoral degrees in zoology, the latter in 1960 on the topic that was later to win him the Nobel, nuclear transplantation in *Xenopus*. For his postdoctoral work, Gurdon then went to the California Institute of Technology (Caltech), a year that turned out to be a substantial adventure for him personally, but a bit of a disappointment, professionally. Determined to make the most of his year in the United States, he flew to New York, where he purchased a second-half Chevrolet in which he drove across the country along the famous Route 66. Along the way, he did a great deal of sightseeing and gave a number of academic lectures on his research at Oxford. At Caltech, however, he was unable to find someone with whom his research interest skills and interests matched, although he later described the experience in an entry for the Ganga Library overall as "extremely formative" because of the introduction it provided to the field of molecular biology.

Upon his return to England, Gurdon was offered a position as assistant lecturer in the department of zoology at Oxford. At first, he was required to give 24 lectures a year, although over the next decade, his work load consisted more and more of research, until his assignment was finally reduced to two lectures per year. In 1971, Gurdon left Oxford to become professor of cell biology at the Medical Research Council Laboratory of Molecular Biology at Cambridge University, where he has held a formal position ever since. In 1983, he became a member of the department of zoology at Cambridge and, in 1989, was a founding member of the Wellcome/CRC (Cancer Research Center) Institute for Cell Biology and Cancer at Cambridge. (In 2004, the institute changed its name to the Wellcome Trust/Cancer Research UK Gurdon Institute, more commonly known simply as The Gurdon Institute.) From 1995 to 2002, Gurdon also served as master of Magdalene College at Cambridge.

In addition to his Nobel Prize, Gurdon has received a number of honors and awards, including the Albert Brachet Prize of the Belgian Royal Academy; Scientific Medal of the Zoological Society; Paul Ehrlich-Ludwig-Darmstaedter Prize; Nessim Habif Prize of the University of Geneva; Ciba Medal and Prize of the Biochemical Society; Priz Charles Leopold Mayer of the Academie des Sciences, France; Ross Harrison Prize of the International Society of Developmental Biology; the Royal Medal of the Royal Society; Emperor Hirohito International Prize for Biology; and Wolf Prize in Medicine. He was elected a fellow of the Royal Society in 1971, made an honorary foreign member of the American Academy of Arts and Sciences in 1978, and a foreign associate of the U.S. National Academy of Sciences in 1980 and the American Philosophical Society in 1983.

Gottlieb Haberlandt (1854–1945)

Haberlandt was an Austrian botanist widely regarded as the founder of plant tissue culture. He was of the opinion that individual tissue taken from various parts of a plant could be cultured and grown into complete plants, a concept now known as totipotentiality. He carried out a number of experiments based on this concept, although none was very successful. His importance in the field, then, was the theoretical basis he provided for the field of tissue culture, which was eventually realized in practice much more successfully by a number of his students and followers, both in the field of plant and animal tissue culture.

Gottlieb Haberlandt was born Ungarisch-Altenburg, Austria, present-day Magyaróvár, Hungary, on November 28, 1854. His father was a professor of botany at the University of Vienna, credited with having made the potential value of soybeans and soyfoods apparent to people in Europe and, later, the United States. His mother was a governess who was, as described by biographer O. Härtel, "well educated and also interested in arts and literature as well as in music." Young Gottlieb apparently inherited an interest in the arts, literature, and music from his mother, but decided to follow in his father's footsteps as a botanist. This decision may have been a result of his early homeschooling by his father, who appears to have imbued his son with a passion for the study of plants.

Haberlandt's early years in Ungarisch-Altenburg were difficult ones for a variety of both personal and political reasons. He was often ill and unable to attend school; at the same time the region was in turmoil as a result of the Austro-Prussian War of 1866, one result of which was the transfer of Ungarisch-Altenburg to Hungary from Austria. Along with other German-speaking residents of the region, Haberlandt and his family found it difficult to adjust to their new Hungarian nationality. In spite of these problems, Haberlandt was able to continue his education at the local government-sponsored agricultural school, where he became qualified as a teacher of botany. He received his first teaching appointment in 1872 as professor of agriculture at the Hochschule für Bodenkulture in Vienna. He held that post for only one year, however, before enrolling at the University of Vienna as a student in botany, where he received his PhD three years later. Härtel reports that Haberlandt chose botany as his major, rather than some other field of biology, because of his aversion to the killing of animals for his research as well as his introduction to an influential textbook on plant physiology by the eminent botanist Julius Sachs, a Christmas present from his father.

Upon completing his doctoral work at Vienna, Haberlandt chose to spend a year working as his father's assistant at Vienna before striking out on his own as a postdoctoral student with Simon Schwendener at the University of Tübingen. At the completion of that year, Haberlandt then traveled to the universities of Berlin and Vienna, where in 1879, he first presented his ideas about the use of plant tissue culture for the study of the form and function of plants. In 1880, he took a position as supplent, or substitute teacher of botany, at the Technical University of Graz. In 1884, Haberlandt was promoted to associate professor at Graz, and four years later, he was appointed full professor botany, succeeding Hubert Leitgeb, who had created the department of botany at Graz, but then committed suicide because so few resources were committed to the successful operation of that department.

Haberlandt remained at Graz until 1910, when he accepted an appointment as professor of botany at the University of Berlin, replacing his former teacher, Schwendener. He remained at Berlin for the remainder of his academic career, retiring in 1923. During his tenure at Berlin, Haberlandt created the Institute for Plant Physiology at Berlin, an institute that was to become an important force in the training of future plant physiologists and the center of research in the field.

Haberlandt's most important book was *Physiologische Pflanzenanatomie (Physiological Plant Anatomy)*, which eventually appeared in six editions and was widely popular also in its English translation. He also wrote a number of other works on plant physiology, one of the most interesting of which was a description of his trip to Java and Ceylon in 1891–1893, *Eine botanische Tropenreise. Indo-malayische Vegetationsbilder und Reiseskizzen* (never translated into English).

In 1881, Haberlandt married a long-time friend, Charlotte Haecker. Their son, Ludwig Haberlandt, continued the family's long tradition of scholarly fame by becoming a reproductive physiologist, sometimes credited as being "the grandfather of the birth control pill." Gottlieb Haberlandt died in Berlin on January 30, 1945.

Hwang Woo-suk (1953–)

For a period of time in 2004 and 2005, Hwang was widely regarded as one of the most successful and influential cloning researchers in the world. In a pair of papers published in the journal *Science* in those two years, he reported that he had been successful in cloning human embryos, from which he was then able to extract human embryonic stem cells. Hwang's research was praised widely by his colleagues and peers because it opened a new door to the cloning of human embryos for therapeutic purposes. He received wild praise from both the South Korean government and the general populace for these supposed discoveries. Only a year after his second paper appeared, however, Hwang was charged with falsifying his research results and charged with embezzlement of government funds and violations of the nation's bioethics laws. In 2007, he was fired from his position at Seoul National University and was banned from conducting further stem cell research and from receiving federal funding for related research.

Hwang Woo-suk was born on January 29, 1953, in the village of Bu-yeo in the province of Chungnam, South Korea. His father died when he was five years old, and he began to work on the family farm to help support his widowed mother and five siblings and to earn enough money to continue his own studies. After graduating from Daejeon High School, he matriculated at Seoul National University, from which he received his bachelor's degree in veterinary medicine in 1977 and his master's degree and doctorate in theriogenology in 1979 and 1982, respectively. (Theriogenology is the study of animal reproduction.) For his postdoctoral studies, Hwang traveled to Hokkaido University in Sapporo, Japan, as a visiting fellow. In 1986, he accepted an offer to join the faculty at Seoul National University, where he remained until being expelled in 2007.

During his years at the National University, Hwang specialized in the reproduction of cattle and became especially interested in artificial reproductive technologies (ARTs), such as in vitro fertilization (IVF) and cloning. In 1993, he announced the birth of the first calf produced by IVF, and later, the first cloned cow (1999), the first cloned pig (2002), and the first cow resistant to so-called mad-cow disease (2003). This research culminated, of course, in his reported (but later refuted) claim of cloning the first human embryo.

The story of Hwang's disgrace appeared to represent the end of his academic and research career. But such has hardly been the case. Even before leaving his post at National University, Hwang had become involved in the creation of a new and private endeavor, the Sooam Biotech Research Foundation, located in Seoul. The purpose of the foundation was to support and carry out research on cloning of a number of animals for a variety of purposes. Over the next decade, Sooam and Hwang have reported a number of significant accomplishments in the field, including the cloning of the first companion dog, Missy, in October 2007; the commercial cloning of the first companion dog, Lancelot, in November 2008; the cloning of dog disease models for diabetes and Alzheimer's disease in 2010; the initiation of a program to clone the extinct woolly mammoth in 2012; and the cloning of a Chinese champion Tibetan mastiff in 2014.

Hwang's future in cloning research is still somewhat uncertain. In a January 2014 article in the journal *Nature*, one science reporter speculated on his chance to regain at least some portion of his past glory. "If the stain [of his earlier dishonor] cannot be washed away," David Cyranoski writes, "perhaps it can be stamped out of memory by hundreds of paws and hooves."

Thomas J. King Jr. (1921–2000)

In 1952, King and his colleague Robert Briggs reported on the successful transplantation of nuclei taking from the blastula cells of a frog into the enucleated oocytes of other frogs of the same species. The procedure marked the successful conduct of an experiment that had been described a decade and half earlier Hans Spemann in his outline of a recommended "fantastical experiment" for producing the clones of animals. Because of this research, King and Briggs are sometimes honored as the fathers of modern cloning research. The procedure they used now forms the basis of the widely employed stem cell nuclear transfer (or transplantation) (SCNT) that is the basis for much modern cloning and stem cell research.

Thomas Joseph King Jr., was born in New York City on June 4, 1921. When his mother died in childbirth, baby Thomas was moved to Ridgefield Park, New York, where he was raised by an aunt. After completing his secondary education, King returned to New York City to matriculate at Fordham University, from which he received his bachelor of science degree in 1943. He then enlisted in the U.S. Army and was assigned as an instructor at the Army Medical Technicians School at Lawson General Hospital in Atlanta. After completing his service in Atlanta, he also served as an officer in the Medical Administrator Corps in the Pacific zone during World War II.

After being discharged from the Army in 1946, King returned once more to New York City, where he enrolled at New York University (NYU) to work on his master's degree. He eventually completed that degree and also earned his PhD in zoology in 1953. While still a graduate student, King served as a teaching fellow at NYU and as an instructor in the department of physiology at Hunter College in New York. In 1950, King was also asked by Briggs to join him as a research fellow at the Institute for Cancer Research (now the Fox Chase Cancer Center) in Philadelphia. It was during this period that Briggs and King completed the research for which they were eventually to become famous. King used that research also as the topic for his doctoral thesis, "The transplantability of nuclei of arrested hybrid blastulae *R. pipiens* × *R. catesbeiana.*"

After his initial appointment at the cancer institute had expired, King decided to stay on at the facility, eventually becoming chair of the Department of Embryology in 1956, replacing his colleague Briggs in that post. He held that position until 1967, after which he left the cancer institute to become professor of biology at Georgetown University in Washington, D.C., where he remained until 1972. He then moved to the National Cancer Institute at Bethesda, Maryland, where he held a variety of administrative posts until 1980. At that point, he returned to Georgetown where he assumed the posts of professor of obstetrics and gynecology and director of the Kennedy Institute of Ethics (until 1983). He later served also as special assistant to the director (until 1988) and deputy director of the Lombardi Cancer Research Center at Georgetown until his retirement in 1990, after which he held the title of deputy director emeritus of the Lombardi Center. King died in Baltimore of cancer on October 25, 2000.

King shared a number of awards and prizes with Briggs for their breakthrough in cloning research, most significant of which was perhaps the Charles Leopold Mayer Prize of the Académie des Sciences in 1972, the highest honor given by the society, and the first time an American had received that important award.

The Long Now Foundation

The Long Now Foundation was established in 1996 to counteract what its founders view as an obsession with short-term thought and action in modern society. They argue that civilization's long-term survival will be better served by focusing on issues, trends, and problems that extend over hundreds or thousands of years, rather than the next year or two, or even the next generation. The first two major projects conceived by the foundation are The Clock and The Library. The Clock is to be a mechanical device that runs with essentially minimal maintenance for a period of at least 10,000 years, powered primarily by solar energy. (For more information about The Clock, see "Time in the 10,000-Year Clock," http://media.longnow. org/files/2/10_AAS_11-665_Hillis.pdf, accessed on January 7, 2015.) The Library Project is a program designed to develop a hand-held guide to more than 1,500 human languages and the connection among them. (For more information about The Library, see "The Rosetta Project," http://rosettaproject.org/ about/, accessed on January 7, 2015.)

Another of the foundation's long-term projects is Revive & Restore, a program to promote and coordinate efforts to clone living examples of extinct species. Its first project is The Great Passenger Pigeon Comeback, a program to produce a living example of a bird that went extinct in 1914. The program consists of a series of steps that is expected to culminate in the use of the genome of the bird's closest living relative, the band-tailed pigeon, as a platform on which to construct the passenger pigeon's own DNA. That DNA will be produced from material available from museum specimens of the extinct bird. The first step in that project, collecting DNA, was completed in October 2013 with the complete sequencing of DNA taken from a specimen known as Passenger Pigeon 1871 from the Royal Ontario Museum. The project currently expects to produce the first living examples of the passenger pigeon in 2022, with a planned date of 2027 for the release of the birds into the wild.

Revive & Restore has established three criteria for the selection of species for de-extinction. First, DNA of less than 500,000 years of age must be available. That criterion would eliminate many extinct species, such as the dinosaurs, from consideration for de-extinction. Second, there must be some living species closely related to the extinct species so that it can be used as a surrogate parent for the extinct species. Third, do current environmental and other conditions exist that would allow the revived species to be returned once more to nature? If not, is there reason to restore a species that can only be bred and maintained in captivity? Finally, are there social, environmental, health, economic, or other reasons for restoring or not restoring the species. Based on these criteria, the foundation has developed a list of other species that might currently be considered for de-extinction. These include the Carolina parakeet, Cuban red macaw, Ivory billed woodpecker, heath hen, dodo, great auk, Easter Island palm, Xerces blue butterfly, quagga, Tasmanian tiger, Madagascar elephant bird, woolly mammoth, Irish elk, and Stellar's sea cow.

Hans Spemann (1869–1941)

Spemann was awarded the 1935 Nobel Prize for Physiology or Medicine for his research on the process by which a differentiated embryo develops out of undifferentiated embryonic cells. The greater part of that research was actually completed by one of Spemann's colleagues, Hilde Mangold, but Mangold was killed in a laboratory accident in 1924 and was therefore not eligible for the Nobel Prize (which can be given only to living individuals). Spemann was famous in his own right for his research on the cloning of amphibians, exploring a wide variety of factors that might be involved in the production of clones from a single embryo.

Hans Spemann was born in Stuttgart, Germany, on June 27, 1869, the eldest of four children born to Johann Wilhelm and Lisinka Hoffman Spemann. The Spemanns were a wealthy and well-educated family that included a number of lawyers and doctors. Hans attended the Eberhard Ludwigs Gymnasium in Stuttgart, where he majored in classical studies, with a view toward becoming a physician. After graduation, he served in the military before working briefly as a bookseller. In 1891, he enrolled as a medical student at the University of Heidelberg, where he fell under the influence of the eminent embryologist Gustav Wolff. At the time, Wolff was engaged in a study of the embryogenesis of newts, attempting to identify the factors that influenced the development of the amphibians. Spemann became interested in this line of research, which was eventually to become the major emphasis of his own academic career.

During a brief period of clinical training at the University of Munich from 1893 to 1894, Spemann decided that he was more interested in research than in clinical work, and he moved to the University of Würzburg to begin a doctoral program in botany, zoology, and physics. He was awarded his PhD by Würzburg in 1895. After a bout with tuberculosis, Spemann returned to Würzburg as a privatdozent (unpaid teacher), where, three years later, he began his historic work on the cloning of salamanders. Spemann remained at Würzburg until 1908, when he accepted an appointment as professor of zoology and comparative anatomy at the University of Rostock. He held that post until 1914, when he moved to the Kaiser Wilhelm Institute of Biology in Berlin, where he became associate director. In 1919, he moved once more, this time to the University of Freiburg-im-Breisgau, where he became professor of zoology, a post he held until his retirement in 1935. Thereafter, he held the title of professor emeritus at Freiburg-im-Breisgau until his death in 1941.

It was at Freiburg-im-Breisgau that Mangold (then Hilde Proescholdt) carried out her groundbreaking work on embryogenesis that was to lead in part to Spemann's Nobel Prize. Her paper was published shortly after her accidental death and is widely to have been thought to be an important element in Spemann's own accomplishments, although he was certainly worthy of the award he was eventually given. Spemann was always generous in acknowledging Mangold's contribution to his own research, referring to some of his most important discoveries as at least partly the result of "The Experiment of Hilde Mangold" in his published papers.

After his retirement from Freiburg-im-Breisgau, Spemann wrote a book describing his long series of experiments, a book for which he is best known today, *Embryonic Development and Induction*. In this book, he describes a "fantastical experiment" that involves the removal of the nucleus from an unfertilized egg and replacing it with a differentiated embryo, an experiment that Spemann himself was unable to carry out, but which is now routinely performed, known as somatic cell nuclear transplantation (SCNT).

James Thomson (1958–)

Thomson is best known for being the first person to isolate and successfully culture human embryonic stem cells. His research was of special significance because, while stem cells from other species had been isolated and cultured, Thomson's work represented the first time that human cells had been made available for extended research. Thomson's laboratory was also involved in some of the earliest research on the development and isolation of human-induced pluripotent stem cells in 2007. Today, Thomson holds joint appointments as director of regenerative biology at the Morgridge Institute for Research, in Madison, Wisconsin; professor of cell and regenerative biology at the University of Wisconsin School of Medicine and Public Health, at Madison; and professor of molecular, cellular, and developmental biology at the University of California, Santa Barbara. He is also chief scientific officer for Cellular Dynamics International, in Madison, a company that produces derivatives of human-induced pluripotent stem cells for drug discovery and toxicity testing.

James Alexander Thomson was born in Oak Park, Illinois, on December 20, 1958. His mother worked as a college administrator, and his father was a certified public accountant. Thomson grew up and attended public schools in Oak Park before matriculating at the University of Illinois at Champagne–Urbana, from which he received his bachelor's degree in biology in 1981. He then continued his education at the Veterinary Medical Scientist Training Program at the University of Pennsylvania, where he was granted his doctorate in veterinary medicine in 1985 and his PhD in molecular biology in 1988. Thomson then pursued his postdoctoral research at the Primate In Vitro Fertilization and Experimental Embryology Laboratory at the Oregon National Primate Research Center from 1989 to 1991. In 1991, he began his residency in veterinary pathology at the Wisconsin Regional Primate Center (WRPC) at University of Wisconsin, Madison. It was at WRPC he began and completed his research on primate and human embryonic stem cells that was to earn him his place in the history of cloning. His first paper on the topic, "Isolation of a Primate Embryonic Stem Cell Line," was published in the August 1995 issue of the *Proceedings of the National Academy of Sciences of the United States of America*. In 1995, Thomson was also appointed chief pathologist at WRPC. Over the next few years, Thomson continued his research on the isolation and culturing of human embryonic stem cells, eventually achieving the success for which he is best known. In 1998, his report on that research first appeared in the journal *Science* under the title "Embryonic Stem Cell Lines Derived from Human Blastocysts."

With his successes in the laboratory, Thomson also solidified his positions in the academic world. In 2007, the year his work on induced pluripotent stem cells was announced, he accepted his current post at the University of California at Santa Barbara. A year later, he was appointed to his other current position as director of regenerative biology at Morgridge. He also holds the title of John D. MacArthur Professor at the University of Wisconsin, Madison.

Thomson has garnered a number of awards and honors for his accomplishments, including the 1999 Gold Plate Award from the American Academy of Achievement, 2002 Lois Pope Annual LIFE International Research Award, Frank Annunzio Award in Science and Technology of the Christopher

Columbus Foundation in 2003, Distinguished Service Award for Enhancing Education through Biological Research of the National Association of Biology Teachers for 2005, Nathan R. Brewer Scientific Achievement Award of the American Association for Laboratory Animal Science in 2006, the Massry Prize of the Meira and Shaul G. Massry Foundation in 2008, a share of the 2011 King Faisal International Prize with Shinya Yamanaka, and the Albany Medical Center Prize in 2013.

Dizhou Tong (1902–1979)

For many people, including most of the population of China who know about cloning, Tong is legitimately entitled to be called the Father of Cloning. During the mid-20th century, he cloned a number of organisms including, in 1963, an Asian carp, at the time, the most complex organisms to have been created by the process of somatic cell nuclear transplantation (SCNT). Tong achieved this step more than three decades before researchers at the Roslin Institute in Scotland cloned the first mammal, the sheep Dolly, for which many historians feel justified in calling Tong the Father of Cloning.

Dizhou Tong (according to Chinese naming custom) was born in a small village in Jing County in Zhejiang Province, People's Republic of China, on May 28, 1902. (He is also known as Ti Chou Tung.) Tong's father was a teacher in his home village and took responsibility for his own son's education at home until he reached the age of 14, when he died and left family responsibilities to his wife. Two years later, Tong enrolled at the Xiaoshi Middle school, where he was then the oldest pupil. After graduating from Xiaoshi in 1923, Tong enrolled at Fudan University in Shanghai, from which he received his degree in biology in 1927.

Upon his graduation from Fudan, Tong followed one of his former teachers, Bao Cai, to his first academic post as assistant in the department of biology at the National Central University in Nanjing. He remained there until 1930, when he decided to continue his studies at the Universite Libre de Bruxelles (Free University of Brussels) in Brussels, Belgium, from which he received his doctorate in biology four years later.

Tong then returned to China where he accepted an appointment as professor of biology at Shandong University in Qhingdao. The beginning of Tong's academic career at Shandong came at a less than fortuitous time only a few years before the outbreak of the disastrous (for China) Second Sino-Japanese War of 1937–1945. During the early years of that war, battles forced the university to move its operations to a number of new locations, and operations were sometimes suspended because of Japanese incursions into an area. Supplies were also difficult to obtain, and it is said that Tong sometimes did not have even the most basic equipment with which to conduct his research. The end of the war in 1945 brought more stability to China and to Tong's own career. In that year, he was promoted to dean of the department of zoology at Shandong where, a year later, he founded the Institute of Marine Biology. The institute later became part of the Chinese Ocean University, located in Qingdao.

In 1948, the U.S. Rockefeller Foundation provided Tong with the funds necessary to spend a year in the United States, during which time he worked as a visiting scholar at Yale University and an independent investigator at the Woods Hole Marine Biological Laboratory. At the end of his sabbatical year in the United States, Tong returned to Shandong, where he was promoted to vice president of the university.

Tong returned to China at a time of renewal and revitalization culminating in the creation of the People's Republic of China in 1949. Among the many administration changes associated with the new government was the creation of the Chinese Academy of Sciences (CAS), which was to become the focus of much of Tong's career for the rest of his life. One of his first acts upon his return to Shandong was the creation of the laboratory of experimental embryology within the Institute of Experimental Biology of the new CAS. He also established a new Laboratory of Marine Biology in Qingdao, which was, within a very short period of time, to become the center of virtually all research in the field in China. Tong was chosen as the first director of the laboratory, later to become the Institute of Oceanology of the CAS, a post he held until his death in 1979. In 1955, Tong resigned his post at Shandong to become head of the Division of Biological Sciences at CAS, a post he also held for more than two decades, eventually becoming vice president of the CAS a year before his death. Throughout his long career of teaching and administration, Tong continued to carry out his research on embryogenesis, including some of the earliest and most successful experiments on somatic cell nuclear transplantation. He died on March 30, 1979.

U.S. Food and Drug Administration

The U.S. Food and Drug Administration (FDA) dates to about 1848, when the U.S. Congress provided for the appointment of Lewis Caleb Beck to the U.S. Patent Office with the charge of carrying out chemical studies of agriculture products. By far the most important enabling legislation related to the FDA, however, was the 1906 Pure Food and Drugs Act, which, for the first time in American history, provided strict regulations about the production, transportation, processing, sale, distribution, and consumption of a wide variety of food products in the United States. The general principle that the U.S. federal government had an essential role to play in protecting the food supply of the American people was laid down in the Pure Food and Drugs Act and, in a sense, continues to act as the guiding principle for today's FDA.

The modern cabinet department, however, has expanded far beyond Lewis Beck's modest beginning a century and a half ago. Today the department has a civilian and military staff of just under 15,000 and an annual budget of about $4.5 billion. The agency's mission is subdivided into eight major categories: food; drugs; medical devices; radiation-emitting products;

vaccines, blood, and biologics; animal and veterinary; cosmetics; and tobacco products. The major divisions through which the FDA carries out its activities are the Office of the Commissioner, which is responsible for most administrative activities and also includes the National Center for Toxicological Research; the Office of Foods and Veterinary Medicine, which includes the Center for Food Safety and Applied Nutrition and the Center for Veterinary Medicine; the Office of Global Regulatory Operations and Policy, which is responsible for international operations; and the Office of Medical Products and Tobacco, which includes offices dealing with biologic research and evaluation, devices and radiologic health, drug evaluation and research, tobacco products, and special medical programs.

The first place to begin in accessing information about the FDA's role in animal cloning policy, regulations, research, evaluation, and education is its Web page on animal cloning at http://www.fda.gov/AnimalVeterinary/SafetyHealth/AnimalCloning/default.htm. This page contains links to the vast majority of important documents produced over the years by various divisions of the FDA on animal cloning. The key FDA agency for dealing with fundamental issues related to animal cloning is the Center for Veterinary Medicine (CVM), which has the primary responsibility for ensuring the safety of foods and drugs used with domestic animals, evaluates the safety of food additives used with animal feed, carries out research on the safety of foods produced from domestic animals, and works toward the development of foods and drugs for so-called minor species, such as fish, hamsters, and parrots.

In 2008, the CVM issued the key document on the safety of food obtained from cloned animals, "Use of Animal Clones and Clone Progeny for Human Food and Animal Feed," available on the FDA Web site at http://www.fda.gov/downloads/AnimalVeterinary/GuidanceComplianceEnforcement/GuidanceforIndustry/UCM052469.pdf. The agency has also issued a large collection of documents that it used in studying this issue and writing the guidelines document. Also available on the FDA Web site are a number of other documents on the general topic of cloning, including a list of consumer FAQs and producer FAQs, records of meetings of important FDA and CVM committees, a primer on cloning with its use in livestock operations and a fact sheet on myths of cloning, a glossary of terms used in cloning research, a backgrounder on genetic engineering in general, and a link to the University of Maryland AgNIC (Agriculture Network Information Center) Agricultural Biotechnology Gateway.

Ian Wilmut (1944–)

The science and technology of cloning has a long and very complex history. It is difficult to point to one specific individual and say that that person was "the father of cloning" or "the most important person" in the field. Yet, one can hardly review the history of cloning without acknowledging the contribution of one person in particular, Ian Wilmut, currently chair of the Scottish Centre for Regenerative Medicine at the University of Edinburgh. Fairly or not, Wilmut is mentioned as the Father of Cloning at least as often as any other person in the history of the field. He has gained this recognition for his role in the cloning of the first mammal, a sheep named Dolly, from an adult somatic cell by nuclear transplantation in 1996. (Somewhat ironically, a dispute has arisen recently as to how important Wilmut's own role was in that research, with he, himself, noting that he had no more than an "administration" or "supervisory" role in the experiments. For more about this dispute, see "Scientists Dispute Credit for Dolly," http://www. theguardian.com/science/2006/mar/11/genetics.highereducation, accessed on January 8, 2015.)

Ian Wilmut was born in Hampton Lucy, Warwickshire, England, on July 7, 1944. His family had been moved from the city of Coventry prior to Ian's birth because of the ongoing bombing of that city by German aircraft toward the end of World War II. Both of Wilmut's parents were teachers, his father a proficient mathematician who had declined an academic career in the field because of his love of teaching. Ian attended the Scarborough Boys High School in northern England, where his father had moved to take a teaching position at that school. But Ian fell under the influence there of an inspiring teacher, Gordon Whalley, who encouraged the young Wilmut to consider biology as a possible career. Some biographers suggest that Wilmut's summer work in the fields around Scarborough may also have convinced him to concentrate in the area of agriculture when he matriculated at the University of Nottingham after completing his secondary education.

In any case, Wilmut received his bachelor's degree from Nottingham in 1967, after which he decided to continue his studies at Darwin College at the University of Cambridge, where he worked in the laboratory of biologist Christopher Polge. Wilmut had spent an eight-week course with Polge prior to his enrolling at Nottingham and decided to continue his affiliation with the eminent researcher. In 1971, Wilmut was awarded his PhD for his research in the field of cryopreservation, Polge's own special field of interest. Wilmut's work in the field of cryopreservation eventually led to his research on the possibility of reproducing animals from frozen gametes and then to his work on the cloning of mammals using the process of somatic cell nuclear transplantation (SCNT). That work, in turn, resulted in the birth of the first cloned mammal from adult somatic cells in 1996.

After receiving his doctorate, Wilmut accepted an appointment at the Roslin Institute, near Edinburgh, then known as the Institute of Animal Physiology and Genetics Research. The institute was reorganized and given its current name in 1993. For his first decade at Roslin, Wilmut's research was focused on the normal process of embryogenesis, with special attention to changes that interfere with that process and result in abnormal forms of development. That line of research was interrupted, however, in 1984 when Wilmut became involved in a project to genetically engineer sheep to develop animals whose milk contained proteins used in the treatment of human diseases. That line of research, in turn, evolved into an interest in using SCNT to reproduce a complete new cloned animal, research that resulted in the birth of Dolly in 196.

Wilmut has continued his affiliation with the Roslin Institute to the present day. He has received a number of awards and honors in recognition of his contribution to cloning research, including the Paul Ehrlich Prize of the Paul Ehrlich Foundation, for outstanding research in the field of medicine, and the Shaw Prize of the Shaw Foundation, for distinguished and significant advances in the field of life science and medicine. Wilmut was also knighted by Queen Elizabeth in 2008 and has been given the Order of the British Empire and elected a fellow of the Royal Society, a fellow of the Royal Society of Edinburgh, and a fellow of the Academy of Medical Sciences of the United Kingdom. In 1997, he was chosen by *Time* magazine as runner-up to its Man of the Year award. Wilmut formally retired in 2011 and now holds the title of Emeritus Professor at the Centre for Regenerative Medicine at the University of Edinburgh. Wilmut is the author of two books, *After Dolly: The Uses and Misuses of Human Cloning* (with Roger Highfield; W. W. Norton, 2006) and *The Second Creation: Dolly and the Age of Biological Control* (with Keith Campbell and Colin Tudge; Farrar, Straus and Giroux, 2000).

Shinya Yamanaka (1962–)

Yamanaka was awarded a share of the 2012 Nobel Prize for Physiology or Medicine for his discovery of a method for producing pluripotent stem cells from fully developed and differentiated adult cells. This discovery was of particular importance because it made possible an alternative to the use of human embryonic stem cells for stem cell research. The discovery of such cells had been demonstrated a decade earlier by two American research teams led by John Thomson at the University of Wisconsin, Madison, and John Gearhart, then at the Johns

Hopkins University in Baltimore, a discovery that had spurred interest enthusiasm for and interest in the use of human stem cells for the treatment of a range of human diseases and disorders. The main roadblock to this line of research, however, was the necessity of creating or using human embryos, which then had to be destroyed in the process of research. The discovery that some adult cells can also be reprogrammed to become pluripotent, as are embryonic stem cells, provided a new approach to stem cell research that did not carry with the ethical implications posed by the use of embryonic stem cells.

Shiya Yamanaka was born on September 4, 1962, in Osaka, Japan. His father, Shozaburo Yamanaka, operated a small factory for the production of sawing machines in Higashi-Osaka, a business with which his mother, Minako, also assisted (in addition to her being a housewife and mother to two children). In his Nobel biography, Yamanaka reports that he was always interested in science as a young boy, usually attempting to follow the line of work pursued by his engineer father. During his school years at Tennoji Junior and Senior High Schools, Yamanaka gradually began to think more seriously of a career in medicine, a career direction encouraged by his father. Yamanaka also reports in his Nobel biography that he was also interested in music in high school, eventually forming his own band, Karesanui (Dry Garden Style), in which he played guitar and was lead singer.

In 1981, Yamanaka enrolled in the School of Medicine at Kobe (Japan) University, from which he received his MD degree in 1987. He then completed his residency at Osaka National Hospital before deciding to continue his studies at the Osaka City University Graduate School, from which he earned his PhD in 1993. Yamanaka then accepted an appointment as postdoctoral fellow (1993–1995) and then staff research investigator (1995–1996) at the Gladstone Institute of Cardiovascular Disease in San Francisco. At the completion of this period, he returned to Japan to become assistant professor at the Osaka City University Medical School from 1996 to 1999, before moving to the Nara Institute of Science and Technology, where he was associate professor from 1999 to 2003 and then professor from 2003 to 2005.

In 2004, Yamanaka was also appointed professor at the Institute for Frontier Medical Sciences at Kyoto University, a post he held until 2010. It was during this period that he published his paper on induced pluripotent stem (iPS) cells mentioned in his 2012 Nobel Prize award, "Induction of Pluripotent Stem Cells from Mouse Embryonic and Adult Fibroblast Cultures by Defined Factors" (with Katzutoshi Takahashi, *Cell*, 126(4): 663–676). From 2007 to 2012, Yamanaka also held the post of professor at the Institute for Integrated Cell-Material Sciences at Kyoto University, and from 2008 to 2010, he was director of the Center for iPS Cell Research and Application at Kyoto. Since 2012, Yamanaka has been professor at the Center for iPS cell Research and Application at Kyoto.

The Nobel Prize was only one of a number of honors accorded to Yamanaka for his work with iPS. He has also received the JSPS Prize of the Japan Society for the Promotion of Science; Asahi Award of the Asahi Newspaper Company; Inoue Prize for Science; Special Prize for Science and Technology of the Japanese Minister of Education, Culture, Sports, Science, and Technology; Shaw Prize in Life Science and Medicine; Medal of Honor with Purple Ribbon of the Japanese government; Robert-Koch Prize for 2008; Canada Gairdner Award of the Gairdner Foundation; Albert Lasker Basic Medical Research Award; Imperial Prize of the Japan Academy; Wolf Prize in Medicine; and Bunka-kunsho (Order of Culture) of the Japanese government.

Introduction

One means of gaining an insight into the history, current status, and controversies associated with hydraulic fracturing is to read through some of the laws, regulations, court cases, speeches, reports, and other documents that have been produced on these topics. A review of relevant data and statistics also provides information on these subjects. This chapter provides excerpts from some of the most important documents dealing with fracking issues as well as basic data and statistics and the use and effects of the procedure.

Data

Table 5.1

National Institutes of Health Stem Cell Research Funding, FY 2002–2013, in millions of dollars.

This table summarizes the amounts allocated by the federal government for various types of stem cell research between 2002 and 2013.

Protesters dressed as cows take a break from marching in a rally to show their opposition to the proposed use of cloned animal products in food in a rally organized by Ben & Jerry's in Washington on February 7, 2007. (AP Photo/Jacquelyn Martin)

Table 5.1 National Institutes of Health Stem Cell Research Funding, FY 2002–2013, in millions of dollars

Year	Human Stem Cells		Nonhuman Stem Cells	
	Embryonic	Nonembryonic	Embryonic	Nonembryonic
2002	10.1	170.9	71.5	134.1
2003	20.3	190.7	113.5	192.1
2004	24.3	203.2	89.3	235.7
2005	39.6	199.4	97.1	273.2
2006	37.8	206.1	10.4	288.7
2007	42.1	203.5	105.9	305.9
2008	88.1	297.2	149.7	497.4
2009 (non-AARA)	119.9	339.3	148.1	550.2
2009 (ARRA)	22.7	57.9	29.1	88.1
2010 (non-ARRA)	125.5	340.8	175.3	569.6
2010 (ARRA)	39.7	73.6	19.6	74.2
2011	123.0	394.6	164.6	619.9
2012	146.5	504.0	163.9	653.0
2013	146.1	431.0	153.7	612.8

*American Recovery and Reinvestment Act

Source: NIH Stem Cell Research Funding, FY 2002–2013. http://stemcells.nih.gov/research/funding/pages/Funding.aspx. Accessed on January 3, 2015.

Table 5.2

The California Institute for Regenerative Medicine (CIRM) was created in 2004 when state voters passed Proposition 71, the California Stem Cell Research and Cures Initiative. That initiative provided for a start-up fund of $3 billion over 10 years to create the institute and fund research on stem cells. This table summarizes the type and amount of grants provided by CIRM through the end of 2012.

Table 5.3

As part of its 2008 study on the safety of food products produced from cloned and noncloned farm animals, the Center for Veterinary Medicine of the U.S. Food and Drug Administration

Table 5.2 Funding for Stem Cell Research, State of California, as of December 2012

Funding Intent	
Training programs	11%
Facilities	20%
Research grants	69%
Stem Cell Use (millions of dollars)	
Embryonic	267
Reprogrammed IPS cell	135
Adult	93
Cancer	16
Other	12
Direct reprogramming	9
SCNT	3
Disease Categories	
Neurological disorders	34%
Heart/vascular disease	15%
Blood/immune disorders	14%
Cancer	12%
Sensory organs	6%
Bone/cartilage disorders	5%
Muscular disorders	5%
Diabetes	3%
Gastrointestinal/liver	3%
Reproductive disorders	2%
Kidney/urinary disorders	1%

Source: Overview of CIRM Stem Cell Research Funding. http://www.cirm.ca.gov/our-progress/stem-cell-research-funding-overview. Accessed on January 3, 2015; Turning Stem Cells into Cures. 2013. California Institute for Regenerative Medicine. http://www.cirm.ca.gov/sites/default/files/files/about_cirm/CIRM_2013_AnnualReport.pdf. Accessed on January 3, 2015.

made a number of comparisons of the composition of food products obtained from both types of animals. Some of the most extensive data used in this analysis were obtained from Cyagra, Inc., a company that had been producing cattle clones since 1999. This table shows the results of that comparison of selected cloned cattle with a sample of noncloned animals.

Table 5.3 Characteristics of Food Products from Cloned and Noncloned Farm Animals

Animal	43	71	72	73	75	78	79	80	119	132	Summary
Sodium	*	*	*	*	*	*	*		*	*	1/10
Potassium	*	*	*	*	*	*	*	*	*	*	0/10
Chloride	#	*	*	*	*	*	*	*	#	*	0/10
Bicarbonate	*	*	*	*	*	*	*	*	*	*	0/10
Anion Gap	*	*	*	*	*	*	*	*	*	*	0/10
Urea	*	*	*	*	*	*	*	*	*	*	0/10
Creatine	*	*	*	*	*	*	*	*	*	*	0/10
Calcium	*	*	*	*	*	*	*	*	*	*	0/10
Phosphate	*	*	*	*	*	*	*	*	*	*	0/10
Magnesium-XB	*	*	*	*	*	*	*	*	*	*	0/10
Total Protein	*	*	*	*	*	*	*	*	*	*	0/10
Albumin-bulk	*	*	*	*	*	*	*	*	*	*	0/10
Globulin	*	*	*	*	*	*	*	*	*	*	0/10
A/G	*	*	*	*	*	*	*	*	*	*	0/10
Glucose	*	#	*	*	#	#	*	*	*	*	0/10
AST/P5P									*		9/10
SDH	*	*	*	*	*	*	*	*	*	*	0/10
Alkaline Phosphate	*	*	*	*	*	*	*	*	*	*	0/10
GGT							*	*	*	*	1/5[1]
Total Bilirubin	*	*	*	*	*	*	*	*	*	*	0/10
Direct Bilirubin	*	*	*	*	*	*	*	*	*	*	0/10
Indirect Bilirubin	*	*	*	*	*	*	*	*	*	*	0/10
Amylase	*	*	*	*	*	*	X	X	*	*	0/8
Cholesterol		#	#			*	*	*	*		4/10
Creatine Kinase	*	*		*	*	*	*	*	*	0	1/10
Iron	*	*	*	*	*	*	*		*		2/10
TIBC	*	#	#		*	#	#	*	*	*	1/10
% Saturation	*	*	*	*	*	*	*		*	*	1/10
hBA-Random						*		*	*	X	6/9
Lipemia	*	*	*	*	*	*	*	*	*	*	0/10
Icterus	*	*	*	*	*	*	*	*	*	*	0/10

Animal	43	71	72	73	75	78	79	80	119	132	Summary
IGF-1	*	*	X	*	*	*	*	*	*	X	0/8
Estradiol	*	*	*	*	*	*	*	*	X	*	0/9
Summary	4/33	3/32	4/32	5/33	4/33	2/33	2/32	4/32	0/32	3/21	

Key:
* No statistical difference
# Statistically different, but clinically irrelevant
↑ Cloned animal greater than comparison animal
↓ Cloned animal less than comparison animal
X No data available
1 As in the original

For explanation of all terms, see Appendix E of the report.

Source: Animal Cloning: A Risk Assessment. Center for Veterinary Medicine, U.S. Food and Drug Administration. January 8, 2008. http://www.fda.gov/downloads/ AnimalVeterinary/SafetyHealth/AnimalCloning/UCM124756.pdf. Accessed on January 14, 2015.

Documents

Dickey Amendment (Public Law 104-99, 1996)

In 1995, the U.S. Congress passed an appropriation bill for the U.S. Department of Health and Human Services, prohibiting the use of federal funds for any research in which human embryos are destroyed or in which such embryos are created for research purposes. The amendment (also known as the Dickey-Wicker Amendment) was to remain the controlling force in U.S. policy for the funding of stem cell research for the better part of two decades. The text of the amendment is (in full) as follows:

(a) None of the funds made available in this Act may be used for—

   (1) the creation of a human embryo or embryos for research purposes; or

   (2) research in which a human embryo or embryos are destroyed, discarded, or knowingly subjected to risk of injury or death greater than that allowed for research on fetuses in utero under 45 CFR 46.208(a)(2) and

section 498(b) of the Public Health Service Act (42 U.S.C. 289g(b)).

(b) For purposes of this section, the term "human embryo or embryos" include any organism, not protected as a human subject under 45 CFR 46 as of the date of the enactment of this Act, that is derived by fertilization, parthenogenesis, cloning, or any other means from one or more human gametes.

(*See* Sherley v. Sibelius *2011, below, for the district and appeals court decisions on the Dickey Amendment.*)

Source: Public Law 104-99. Section 128. *U.S. Statutes at Large*, 110 (1997): 34.

European Union Resolution on Human Cloning (1998)

The European Union has discussed and adopted a number of resolutions on cloning issues. One of its earliest statements on the question of human cloning is the following, adopted in 1998.
The European Parliament,

– having regard to its resolution of 12 March 1997 on cloning animals and human beings ((OJ C 115, 14.4.1997, p. 92.)),

– having regard to its opinion of 16 July 1997 on the proposal for a European Parliament and Council Directive on the legal protection of biotechnological inventions ((OJ C 286, 22.9.1997, p. 87.)),

– having regard to the 1996 Council of Europe Convention for the Protection of Human Rights and Dignity of the Human Being with regard to the Application of Biology and Medicine, also known as the 'human rights and biomedicine convention', and the additional protocol prohibiting human cloning,

A. having regard to the disquiet caused by the announcement by an American researcher of his intention to clone human beings,

B. whereas human cloning is defined as the creation of human embryos having the same genetic make-up as another human being, dead or alive, at any stage of its development from the moment of fertilization, without any possible distinction as regards the method used,

C. whereas the cloning of human beings, whether carried out on an experimental basis, in the context of fertility treatments, preimplantation diagnosis, for tissue transplantation, or for any other purpose whatsoever, is unethical, morally repugnant, contrary to respect for the person and a grave violation of fundamental human rights which cannot under any circumstance be justified or accepted,

D. whereas scientific research, which is one of the keys to human progress, must be pursued; whereas, however, it may not undermine the dignity and integrity of the human being,

E. whereas on 12 January 1998, the Council of Europe opened for signature the first and only international text imposing a stringent ban on human cloning,

1. Reiterates that every individual has the right to his own genetic identity and that human cloning must be prohibited;

2. Calls on the Member States of the Council of Europe to sign and ratify the Council of Europe human rights and biomedicine convention and its additional protocol prohibiting human cloning, as these two legal instruments are binding and provide for severe penalties in the event of their being breached in the countries of Europe; in the event of a difference of opinion on the definition, calls for the definition given by the European Parliament to be used;

3. Calls on each Member State to enact binding legislation prohibiting all research on human cloning

within its territory and providing for criminal sanctions for any breach;

4. Calls on the Member States, the European Union and the United Nations to take all the steps necessary to bring about a universal and specific ban, which is legally binding, on the cloning of human beings, including the convening of a world conference on this subject;

5. Calls on the international scientific community, in carrying out research on the human genome, to refrain from the cloning of human beings;

6. Reminds the Council of Parliament's insistence that no Community funds should be used, directly or indirectly, for research programmes which make use of human cloning and calls for confirmation that this prohibition is being fully applied;

7. Instructs its President to forward this resolution to the Commission, the Council, the governments of the Member States, the Secretary-General and the Parliamentary Assembly of the Council of Europe and the Secretary-General of the United Nations.

Source: Resolution on Human Cloning. 1998. http://eur-lex .europa.eu/legal-content/EN/TXT/HTML/?uri=CELEX:519 98IP0050&rid=13. Accessed on January 3, 2015.

Issues Raised by Human Cloning Research (2001)

On March 28, 2001, the Subcommittee on Oversight and Investigations of the Committee on Energy and Commerce of the House of Representatives held a hearing on Issues Raised by Human Cloning Research. The committee received written and oral testimony from a number of individuals speaking in support of and in opposition to research on human reproductive cloning. One of the participants was Raël, whose birth name was Claude Maurice Marcel Vorilhon, the founder of the Raëlian religion, which teaches that the human race is descended from an alien civilization. In late 2002, Raëlians announced that they had cloned the first human child, a claim that was never proved. In his testimony before the House subcommittee, Raël explained his reasons for supporting research on human reproductive cloning and arguing that no additional committees on the ethics of the practice were needed. (All punctuation and grammar as in the original.).

The conservative, orthodox, fanatic traditional religions have always tried to keep humanity in a primitive stage of darkness. It is easy to see that in countries like Afghanistan which are back to the middle ages due to a fanatic Moslem government.

But this was also true in occidental powers. The first medical doctors who tried to study the human by opening cadavers were excommunicated by the Catholic Church. It was considered a sin to try to unveil the mystery of the creation of god. . . . So were the first antibiotics, blood transfusions, vaccines, surgeries, contraception, organ transplants . . . religious fanatics were always saying that "it's a sin to go against the will of god. . . . If somebody is sick, let him die, his life is in god's hands."

If our civilization would have respected these primitive ideas from dark ages, we would all die around 35, and 9 out of 10 babies born would die in their first 2 years.

Traditional religions have always been against scientific progress. They were against the steam engine, electricity, airplanes, cars, radio, television, etc. . . .

If we had listened to them we would still have horses and carts and candles. . . . Twenty-two years ago they were against IVF, talking about monsters, Frankenstein and playing god, and now IVF is well accepted, performed every day by thousands and helping happy families with fertility problems to have babies.

Today human cloning will help other families to have children, and again they are against it. It will also help to cure numerous diseases, will help us live a lot longer, and finally will help us reach, in the future, eternal life.

Nothing should stop science, which should be 100% free.

Ethical committees are unnecessary and dangerous as they give power to conservative, obscurantist forces, which are guided only by traditional religious powers.

As well as there should be a complete separation of state and religion, there should also be a complete separation of science and state, or science and religion.

If there was an ethical committee when antibiotics, blood transfusions and vaccines were discovered it would have certainly been possible that these technologies would have been forbidden. You can imagine the poor health the world would be in today. . . .

Ethical committees should be necessary when a deadly technology is making the production of weapons of mass destruction possible. . . . And to my knowledge there are no ethical committees concerning nuclear, chemical or biological weapons. These things are created to kill millions of people and possibly destroy all life on earth. Cloning is a pro-life technology, a technology made to give birth to babies!

The first benefit of human cloning is to make it possible for couples who cannot have children using other existing techniques to have babies inheriting genetic traits from one of their parents. They can be unfertile heterosexual couples or gay couples.

The second benefit is for families who lose a child due to crime, accident or disease to have the same child brought back to life.

All conservative "pro life" groups always talk about "the right of the unborn", but in this case we must talk about protecting the rights of the "unreborn". As cloning technology makes this possible, why should we accept the accidental death of a beloved child, when we can bring this very child back to life?

People who are opposed to it are always influenced by a terrible Judeo-Christian education . . . the same as those who could have made antibiotics, vaccines, transfusions, surgery and organ transplants forbidden.

Source: Issues Raised by Human Cloning Research. 2001. http://www.gpo.gov/fdsys/pkg/CHRG-107hhrg71495/html/ CHRG-107hhrg71495.htm. Accessed on January 5, 2014.

Benefits of Human Reproductive Cloning (2002)

While cloning is certainly a controversial topic, arguments in favor of human reproductive cloning tend to be relatively uncommon. An organization called the Human Cloning Foundation, however, long promoted this form of cloning. In 2002, it presented an article on its Web site outlining what it saw as the benefits of human reproductive cloning. It also listed a number of benefits that, it said, had been mentioned in emails to the organization. Among those benefits were the following:

1) A couple has one child then they become infertile and cannot have more children. Cloning would enable such a couple to have a second child, perhaps a younger twin of the child they already have.

2) A child is lost soon after birth to a tragic accident. Many parents have written the HCF after losing a baby in a fire, car accident, or other unavoidable disaster. These grief stricken parents often say that they would like to have their perfect baby back. Human cloning would allow such parents to have a twin of their lost baby, but it would be like other twins, a unique individual and not a carbon copy of the child that was lost under heartbreaking circumstances.

3) A woman who through some medical emergency ended up having a hysterectomy before being married or having children. Such women have been stripped of their ability to have children. These women need a surrogate mother to have a child of their own DNA, which can be done either by human cloning or by in vitro fertilization.

4) A boy graduates from high school at age 18. He goes to a pool party to celebrate. He confuses the deep end and shallow end and dives head first into the pool, breaking his

neck and becoming a quadriplegic. At age 19 he has his first urinary tract infection because of an indwelling urinary catheter and continues to suffer from them the rest of his life. At age 20 he comes down with herpes zoster of the trigeminal nerve. He suffers chronic unbearable pain. At age 21 he inherits a 10 million dollar trust fund. He never marries or has children. At age 40 after hearing about Dolly being a clone, he changes his will and has his DNA stored for future human cloning. His future mother will be awarded one million dollars to have him and raise him. His DNA clone will inherit a trust fund. He leaves five million to spinal cord research. He dies feeling that although he was robbed of normal life, his twin/clone will lead a better life.

5) Two parents have a baby boy. Unfortunately the baby has muscular dystrophy. They have another child and it's another boy with muscular dystrophy. They decide not to have any more children. Each boy has over 20 operations as doctors attempt to keep them healthy and mobile. Both boys die as teenagers. The childless parents donate their estate to curing muscular dystrophy and to having their boys cloned when medical science advances enough so that their DNA can live again, but free of muscular dystrophy.

Source: The Benefits of Human Cloning. HumanCloning. org. http://www.humancloning.org/benefits.php. Accessed on January 2, 2015.

Therapeutic versus Reproductive Cloning (2002)

The debate over reproductive versus therapeutic cloning is a long and contentious one. Many thoughtful arguments have been presented in favor of or in opposition to therapeutic cloning, although the vast majority of observers have opposed human reproductive cloning under virtually all circumstances. One of the most cogent arguments in support of therapeutic cloning was one presented at a 2002 hearing before the Committee on the Judiciary of the U.S. Senate. That argument is presented in its entirety here.

The proponents of the bans [against all forms of cloning], for example, fear that once a cloned embryo exists in a laboratory, either the embryo or its so-called parent may have constitutional grounds to insist that pregnancy be permitted. But this makes no sense. It requires either that the embryo itself have a constitutional right to be born—something that the U.S. Supreme Court has specifically rejected and also has been rejected by leading State courts hearing disputes over existing frozen IVF embryos in our laboratory now—or this argument would require that people be considered to have a fundamental right to use these embryos to reproduce through cloning—in other words, to have a fundamental right to reproduce by cloning, per se, which would render the entire ban equally unconstitutional.

Now, others worry not about a constitutional ground for bringing the embryo to term but simply that the cloned embryo sitting in a lab will tempt someone to use it illegally. But I would note that the criminal penalties in bills such as S. 1758 are equally effective whether the cloned embryo already exists or is merely imagined. The deterrent is clear, and it is not strengthened by criminalizing basic research.

So if criminalizing research is not needed to guard against the unfortunate outcome of using cloning to make children, it must have another purpose. And indeed the proponents have cited the research ban being needed to protect embryos, women's health, and even the future of humanity.

In my opinion, if the purpose is to protect embryos, then criminalizing research and so-called therapeutic cloning is an odd place to begin. As Senator Durbin has already pointed out this afternoon, we know and, indeed, we fully expect that embryos will unfortunately be lost by the thousands each year at in vitro fertilization clinics, even if IVF is done perfectly, even if every woman who wants to adopt an embryo is successful.

Criminalizing research cannot alter the scale of this embryo loss, and since almost no one thinks that IVF itself could be outlawed, then banning a technique that might involve an exceedingly small number of embryos represents, at best, a symbolic effort at embryo protection.

Now, such symbolic efforts are important. They remind us that life is a gift to be experienced with awe and gratitude. But such symbols can be badly tarnished if they are adopted at the expense of pain and suffering. And as Dr. Weissman has noted and as the chairman noted when she first opened the hearing, reproductive cloning at this time is a danger to children but non-reproductive cloning might save their lives. Whether by doing research with the cells of those who have genetic diseases so we can study in a laboratory dish how the defective gene operates and develop drugs to treat, or to use it for transplantation without risk of rejection, it is potentially life-saving. But most important—and, again, as the chairman noted in her earlier remarks—studying research cloning allows us to understand how cloning reprograms adult cells, which may in the future allow us to reprogram those cells directly without cloning and without the use of embryos in order to generate tissue that could be used to alleviate paralysis or save lives.

Yes, there are other promising avenues of research, and you will hear about them this afternoon. They most certainly should be pursued. But that is no argument for criminalizing this research.

America is not a country in which basic research or personal choices are illegal until someone has persuaded the government to grant permission. Quite the contrary. We celebrate the freedom to think and to act and to inquire into the secrets of nature until a compelling case can be made that it must be stopped. Identifying complementary areas of research falls far short of making that case. In my opinion, at best it is an argument for shaping Federal funding priorities in a way that affords these alternative avenues every chance of success.

In my last remaining seconds, I would like to note that there are a handful of women's health and environmental organizations long known for a particularly great skepticism about medical science and biotechnology that have also testified against research cloning, saying that it is the first step on a slippery slope toward eugenics and the commodification of life. I would say that therapeutic cloning and research cloning is neither the beginning nor the end of that slippery slope, nor is it even the most important landmark.

Our power over human reproduction is as old as ancient contraceptive potions, and it was IVF that was the true landmark moment at which we were able to manipulate the embryo because it now existed outside the body.

By contrast, cloning research does not engineer or design the embryo, and, indeed, precisely because it does not involve making babies, it does not design or engineer our children. It is not basic research but, rather, our choices about its applications that will shape the future.

A moratorium on attempting pregnancy with cloned embryos is an effective and excellent speed bump on the slippery slope toward this future so many seem to predict and fear. To ask for more and to halt such basic research is to sacrifice the diabetic children, paralyzed police officers, and declining elderly of the present for a future that is neither certain nor imminent.

Criminalizing research cloning is not the way to protect embryos. It is not the way to guard against the future. It merely gambles with the hope held by many people today that they may live to see that future, whatever it may hold. Thank you very much.

Source: Statement of R. Alta Charo, Professor of Law and Medical Ethics, University of Wisconsin Law School, Madison, Wisconsin. 2002. "Human Cloning: Must We Sacrifice Medical Research in the Name of a Total Ban?" Hearing before the Committee on the Judiciary of the U.S. Senate. http://www.gpo.gov/fdsys/pkg/CHRG-107shrg83684/pdf/CHRG-107shrg83684.pdf. Accessed on January 6, 2015.

North Dakota Law on Cloning: Chapter 12.1–39: Human Cloning (2004)

As of early 2015, 18 states have laws regulating human cloning. Some laws prohibit both therapeutic and reproductive cloning, whereas others prohibit only the latter. Six states also prohibit cloning for any type of research purpose. One of the simplest of these laws is North Dakota's, of which the second section is quoted here in full. The first section provides definitions for terms used in the law.

12.1–39–02 Human Cloning—Prohibition—Penalty

1. A person may not intentionally or knowingly:

   a. Perform or attempt to perform human cloning:

   b. Participate in performing or attempting to perform human cloning:

   c. Transfer or receive the product of a human cloning for any purpose; or

   d. Transfer or receive, in whole or in part, any oocyte, human embryo, human fetus, or human somatic cell, for the purpose of human cloning.

2. Nothing in subsection 1 restricts areas of scientific research not specifically prohibited, including in vitro fertilization, the administration of fertility-enhancing drugs, or research in the use of nuclear transfer or other cloning techniques to produce molecules, deoxyribonucleic acid, tissues, organs, plants, animals other than humans, or cells other than human embryos.

3. A person who violates subdivision a or b of subsection 1 is guilty of a class C felony. A person who violates subdivision c or d of subsection 1 is guilty of a class A misdemeanor.

Source: Chapter 21–1.39. Human Cloning. North Dakota CenturyCode. http://www.legis.nd.gov/cencode/t12-1c39.pdf? 20140726163223. Accessed on July 26, 2014.

United Nations Resolution on Human Cloning, 59/280 (2005)

On March 8, 2005, the U.N. General Assembly adopted the following resolution regarding the cloning of human beings.

United Nations Declaration on Human Cloning
  The General Assembly,

> *Guided* by the purposes and principles of the Charter of the United Nations,

> *Recalling* the Universal Declaration on the Human Genome and Human Rights, adopted by the General Conference of the United Nations Educational, Scientific and Cultural Organization on 11 November 1997,1 and in particular article 11 thereof, which states that practices which are contrary to human dignity, such as the reproductive cloning of human beings, shall not be permitted,

> *Recalling* also its resolution 53/152 of 9 December 1998, by which it endorsed the Universal Declaration on the Human Genome and Human Rights,

> *Aware* of the ethical concerns that certain applications of rapidly developing life sciences may raise with regard to human dignity, human rights and the fundamental freedoms of individuals,

> *Reaffirming* that the application of life sciences should seek to offer relief from suffering and improve the health of individuals and humankind as a whole,

> *Emphasizing* that the promotion of scientific and technical progress in life sciences should be sought in a manner that safeguards respect for human rights and the benefit of all,

> *Mindful* of the serious medical, physical, psychological and social dangers that human cloning may imply for the individuals involved, and also conscious of the need to prevent the exploitation of women,

> *Convinced* of the urgency of preventing the potential dangers of human cloning to human dignity,

Solemnly declares the following:

(a) Member States are called upon to adopt all measures necessary to protect adequately human life in the application of life sciences;

(b) Member States are called upon to prohibit all forms of human cloning inasmuch as they are incompatible with human dignity and the protection of human life;

(c) Member States are further called upon to adopt the measures necessary to prohibit the application of genetic engineering techniques that may be contrary to human dignity;

(d) Member States are called upon to take measures to prevent the exploitation of women in the application of life sciences;

(e) Member States are also called upon to adopt and implement without delay national legislation to bring into effect paragraphs (a) to (d);

(f) Member States are further called upon, in their financing of medical research, including of life sciences, to take into account the pressing global issues such as HIV/AIDS, tuberculosis and malaria, which affect in particular the developing countries.

Source: Resolution adopted by the General Assembly on 8 March 2005. 2005. http://www.un.org/en/ga/search/view_doc.asp?symbol=A/RES/59/280. Accessed on January 6, 2015. Reproduced with kind permission of the United Nations.

Maryland Stem Cell Research Fund (2006)

Another type of state legislation dealing with cloning (see North Dakota 2004 and Montana 2009) goes beyond the prohibition of various types of cloning and specifically provides for the encouragement of one type of cloning—stem cell research—in the state. A law passed in the state of Maryland is an example of this type of legislative action, with similar laws having been adopted in five states as of 2015.

Act of 2006

§ 10-430. Definitions.
This section provides definitions for terms used in the act.

§ 10-430. Creation of Stem Cell Lines Not Prohibited.
Nothing in this part may be construed to prohibit the creation of stem cell lines to be used for therapeutic research purposes.

§ 10-431. Commission Established.
This section defines the membership of the Stem Cell Research Commission.

§ 10-432. Commission Duties.

§ 10-433. Repealed.

§ 10-434. Establishment and Purpose of Fund.

(a) There is a Maryland Stem Cell Research Fund.
(b) The purpose of the Fund is to promote State-funded stem cell research and cures through grants and loans to public and private entities in the State.
(c) The Corporation shall administer the Fund.
(d)

   (1) The Fund is a special, nonlapsing fund that is not subject to reversion under § 7-302 of the State Finance and Procurement Article.

   (2) The Treasurer shall hold the Fund separately, and the Comptroller shall account for the Fund.

(e) The Fund consists of:

   (1) appropriations as provided in the State budget; and

(2) any other money from any other source accepted for the benefit of the Fund.

(f) Money in the Fund may only be used to:

(1) award grants and loans for State-funded stem cell research, in accordance with the recommendations of the Commission;

(2) award grants and loans for facilities, capital leases, and capital equipment where State-funded stem cell research is conducted, in accordance with the recommendations of the Commission; and

(3) pay the costs necessary to administer the Fund.

(g)

(1) The Treasurer shall invest the money in the Fund in the same manner as other State money may be invested.

(2) Any investment earnings shall be paid into the Fund.

(h)

(1) The Governor may include in the State budget bill each fiscal year an appropriation to the Fund.

(2) Expenditures from the Fund may only be made in accordance with an appropriation approved by the General Assembly in the State budget or by an approved budget amendment.

§ 10-435. Grant and Loan Contingencies.
This section provides details of the awarding of grants.

§ 10-436. Independent Scientific Peer Review Committee.

(a) The Commission shall contract with an independent scientific peer review committee composed of scientifically recognized experts in the field of stem cell research.

(b) The committee shall:

(1) review, evaluate, rank, and rate research proposals for State-funded stem cell research:

    (i) based on the procedures and guidelines established by the Commission; and

    (ii) in a manner that gives due consideration to the scientific, medical, and ethical implications of the research; and

(2) make recommendations to the Commission, based on the rankings and ratings awarded to research proposals by the committee, for the award and disbursement of grants and loans under the Fund.

(c) A member of the committee:

    (1) is not eligible to receive a grant or loan for State-funded stem cell research from the Fund;

    (2) may not reside in the State; and

    (3) shall be subject to conflict of interest standards that are at least as stringent as the standards on conflict of interest adopted by the National Institutes of Health.

§ 10-437. Considerations.

(a) A person who conducts State-funded stem cell research shall conduct the research in a manner that considers the ethical and medical implications of the research.

(b) A person who conducts State-funded stem cell research may not engage in any research that intentionally and directly leads to human cloning.

§ 10-438. Health Care Practitioner — Unused Material.

(a) A health care practitioner licensed under the Health Occupations Article who treats individuals for infertility shall:

    (1) provide individuals with information sufficient to enable them to make an informed and voluntary choice regarding the disposition of any unused material; and

    (2) present to individuals the option of:

        (i) storing or discarding any unused material;

        (ii) donating any unused material for clinical purposes in the treatment of infertility;

        (iii) except as provided in subsection (b) of this section, donating any unused material for research purposes; and

        (iv) donating any unused material for adoption purposes.

(b) Any unused material donated for State-funded stem cell research may not be an oocyte.

(c) An individual who donates any unused material for research purposes under subsection (a)(2) of this section shall provide the health care practitioner with written consent for the donation.

§ 10-439. Restrictions and Penalties of Donated Material.

§ 10-440. Prohibition on Human Cloning.

(a) A person may not conduct or attempt to conduct human cloning.

(b) A person who violates this section is guilty of a felony and on conviction is subject to imprisonment not exceeding 10 years or a fine not exceeding $200,000 or both.

§ 10-441. Adoption of Regulations.

§ 10-442. Annual Report.

Source: Act of 2006. Maryland Stem Cell Research Fund. http://www.mscrf.org/content/aboutus/actof2006.cfm. Accessed on July 31, 2014.

Cloned Food Labeling Act, H.R. 992, 110th Congress (2007)

Two separate, but related, issues often arise in discussions about the production and sale of foods that have been genetically modified. The first question is whether or not companies should even be allowed to produce such foods. The second question is, if such foods are permitted, should they be required to be labeled so that consumers are aware that they are purchasing and possibly consuming food products from cloned plants and animals. A number of laws have been proposed to deal with the latter issue, an example of which is House bill 992, introduced by Representative Rosa DeLauro (D-CT) in 2007. The main thrust of that bill is as follows (no action was ever taken on the bill):

A Bill

To amend the Federal Food, Drug, and Cosmetic Act and the Federal Meat Inspection Act to require that food that contains product from a cloned animal be labeled accordingly, and for other purposes.

1. Short title

This Act may be cited as the Cloned Food Labeling Act.

2. Amendments to the Federal Food, Drug, and Cosmetic Act

(a) In general

> Section 403 of the Federal Food, Drug, and Cosmetic Act (21 U.S.C. 343) is amended by adding at the end the following:
> (z)

(1) If it contains cloned product unless it bears a label that provides notice in accordance with the following:
   (A) A notice as follows: THIS PRODUCT IS FROM A CLONED ANIMAL OR ITS PROGENY.

(B) The notice required in clause (A) is of the same size as would apply if the notice provided nutrition information that is required in paragraph (q)(1).

(C) The notice required under clause (A) is clearly legible and conspicuous.

(2) For purposes of this paragraph:

(A) The term cloned animal means—

(i) an animal produced as the result of somatic cell nuclear transfer; and

(ii) the progeny of such an animal.

(B) The term cloned product means a product or byproduct derived from or containing any part of a cloned animal.

(3) This paragraph does not apply to food that is a medical food as defined in section 5(b) of the Orphan Drug Act.

(4)

(A) The Secretary, in consultation with the Secretary of Agriculture, shall require that any person that prepares, stores, handles, or distributes a cloned product for retail sale maintain a verifiable recordkeeping audit trail that will permit the Secretary to verify compliance with this paragraph and subsection (aa).

(B) The Secretary, in consultation with the Secretary of Agriculture, shall publish in the Federal Register the procedures established by such Secretaries to verify compliance with the recordkeeping audit trail system required under clause (A).

(C) The Secretary, in consultation with the Secretary of Agriculture, shall, on annual basis, submit to Congress a report that describes the progress and activities of the recordkeeping audit trail system and compliance verification procedures required under this subparagraph.

Source: Text of the Cloned Food Labeling Act. https://www.govtrack.us/congress/bills/110/hr992/text. Accessed on January 14, 2015.

Guidance for Industry: Use of Animal Clones and Clone Progeny for Human Food and Animal Feed (2008)

In January 2008, the U.S. Food and Drug Administration (FDA) published a document dealing with the safety of food produced from cloned animals. The major conclusions from that document were as follows:

IV. Human Food Derived from Clones

No unique risks for human food consumption were identified in cattle, swine, or goat clones derived via SCNT. No anomalies have been observed in animals produced by cloning that are not also observed in animals produced by other assisted reproductive technologies (ARTs) and natural mating. The frequency of those anomalies, however, is increased over other ARTs and natural mating. As was the case with other ARTs, the success rate is improving over time. Further, the results of the Risk Assessment have clearly indicated that cloning falls on the continuum of assisted reproductive technologies (ARTs).

Following extensive review, the Risk Assessment did not identify any unique risks for human food from cattle, swine, or goat clones, and concluded that there is sufficient information to determine that food from cattle, swine, and goat clones is as safe to eat as that from their more conventionally-bred counterparts. Because of these reasons, and because food from clones would be subject to the same requirements as food from their conventionally bred counterparts, we do not believe that meat or milk from cattle, swine, and goat clones would require any additional controls compared with meat or milk from cattle, swine, or goats currently entering the food supply today.

As stated in the Risk Assessment, insufficient information was available on sheep clones to make a decision on food consumption risks and assessments were not conducted for animals other than cattle, swine, goat, and sheep. Therefore, at this time, the agency continues to recommend that edible products from clones from animals other than cattle, swine, or goat (e.g., sheep) not be introduced into the human food supply.

Source: "Guidance for Industry: Use of Animal Clones and Clone Progeny for Human Food and Animal Feed." U.S. Food and Drug Administration. January 15, 2008. http://www.fda.gov/downloads/AnimalVeterinary/GuidanceComplianceEnforcement/GuidanceforIndustry/UCM052469.pdf. Accessed on July 26, 2014.

European Union Policy on Cloning of Animals for Food Supply (2008)

After an extended period of discussion, deliberation, and debate, the European Parliament in 2008 adopted the following resolution regarding the cloning of animals for food products.

The European Parliament,
   —having regard to Rule 108(5) of its Rules of Procedure,

A. whereas the Protocol on protection and welfare of animals requires the Community and Member States to pay full regard to animal welfare requirements in formulating and implementing agriculture and research policies,
B. whereas cloning processes show low rates of survival for transferred embryos and cloned animals, with many cloned animals dying in the early stages of life from cardiovascular failure, immuno-deficiencies, liver failure, respiratory problems, and kidney and musculoskeletal abnormalities,
C. whereas the European Food Safety Authority (EFSA) concluded in its opinion of 2008 that mortality and morbidity levels in cloned animals are higher than in sexually produced animals and late pregnancy losses and disorders are likely to affect the welfare of surrogate mothers,

D. whereas, given current levels of suffering and health problems of surrogate dams and cloned animals, the European Group on Ethics in Science and New Technologies (EGE) questions whether cloning animals for food supply is ethically justified and does not view as convincing arguments to justify food production from cloned animals and their offspring,

E. whereas Council Directive 98/58/EC of 20 July 1998 concerning the protection of animals kept for farming purposes (1) provides that 'natural or artificial breeding or breeding procedures which cause or are likely to cause suffering or injury to any of the animals concerned must not be practised' (Annex, paragraph 20),

F. whereas cloning would significantly reduce genetic diversity within li livestock populations, increasing the possibility of whole herds being wiped out by diseases to which they are susceptible,

G. whereas EFSA published on 24 July 2008 a scientific opinion on the implications of animal cloning for food safety, animal health and welfare and the environment, in which it concluded that the health and welfare of a significant proportion of cloned animals was adversely affected, often severely and fatally,

H. whereas, while the principal purpose of cloning is to produce multiple copies of animals with fast growth rates or high yields, traditional selective breeding has already led to leg disorders and cardiovascular malfunction in fast-growing pigs, and lameness, mastitis and premature culling in high-yielding cattle; and whereas cloning the fastest-growing and highest-yielding animals will lead to even higher levels of health and welfare problems,

I. whereas, in addition to the fact that the implications of the cloning of animals for food supply have not been adequately studied, it poses a serious threat to the image and substance of the European agricultural model, which is based on product quality, environment-friendly principles and respect for stringent animal welfare conditions,

1. Calls on the Commission to submit proposals prohibiting for food supply purposes (i) the cloning of animals, (ii) the farming of cloned animals or their offspring, (iii) the placing on the market of meat or dairy products derived from cloned animals or their offspring and (iv) the importing of cloned animals, their offspring, semen and embryos from cloned animals or their offspring, and meat or dairy products derived from cloned animals or their offspring, taking into account the recommendations of EFSA and the EGE;

2. Instructs its President to forward this resolution to the Council and the Commission.

Source: European Parliament. http://eur-lex.europa.eu/legal-content/EN/TXT/PDF/?uri=CELEX:52008IP0400&rid=2. Accessed on January 3, 2015. Reproduced with permission of the European Parliament.

Montana Law on Cloning (2009)

In contrast to the type of cloning law adopted in North Dakota (see earlier document), some states have passed laws that prohibit reproductive cloning, but that allow—either expressly or by inference—therapeutic cloning. An example of this type of law is one passed in Montana in 2009. The full text of that law is as follows:

50–11–102. Prohibited acts—penalties. (1) A person or entity, public or private, may not knowingly:

(a) perform or attempt to perform reproductive human cloning;
(b) participate in an attempt to perform reproductive human cloning;

(c) ship, transfer, or receive for any purpose an embryo for reproductive human cloning; or

(d) ship, transfer, or receive, in whole or in part, any oocyte,
embryo, fetus, or human somatic cell for the purpose of
reproductive human cloning.

(2) A violation of subsection (1)(a) or (1)(b), or both, is a
felony.

(3) A violation of subsection (1)(c) or (1)(d), or both, is a
misdemeanor.

(4) All fines collected under this section must be deposited in
the state general fund.

Source: Montana Code Annotated 2013. http://leg.mt.gov/
bills/mca/50/11/50-11-102.htm. Accessed on July 31, 2014.

Sherley v. Sebelius **(704 F. Supp. 2d 63, 2010)**

*In 2009, two stem cell researchers, James L. Sherley and Theresa
Deisher, sued the Secretary of Health and Human Services, Kath-
leen Sebelius, to prevent the Department of Health and Human
Services (HHS) from continuing to fund embryonic stem cell research under President Barack Obama's 2009 executive order on
the issue. The researchers, whose own work used only adult stem
cells, argued that Obama's executive order and existing HEW policy violated the Dickey Amendment, which had banned embryonic
stem cell research for nearly two decades (see 1995). Judge Royce C.
Lamberth granted a preliminary injunction on the research, agree-
ing with the plaintiffs' arguments in the case. He first noted that:*

In 1999, defendants determined that the Dickey-Wicker
Amendment was not applicable to ESC research because ESCs
[embryonic stem cells] are not embryos as defined by statute.
(See Lingo Decl. Ex. D.) Specifically, defendants recognized a
distinction between deriving ESCs from an embryo, which is
prohibited by the Dickey-Wicker Amendment because it results in the destruction of the embryo, and research on ESCs,
which does not result in the destruction of an embryo.

Lamberth then explained his disagreement with this position:

Defendants' argument is unavailing. Their entire argument assumes that the Dickey-Wicker Amendment is ambiguous and that, as a result, they are entitled to *Chevron* deference. As discussed above, defendants' assumption is incorrect. The Dickey-Wicker Amendment unambiguously prohibits the use of federal funds for all research in which a human embryo is destroyed. It is not limited to prohibit federal funding of only the "piece of research" in which an embryo is destroyed. Thus, if ESC research is research in which an embryo is destroyed, the Guidelines, by funding ESC research, violate the Dickey-Wicker Amendment.

ESC research is clearly research in which an embryo is destroyed. To conduct ESC research, ESCs must be derived from an embryo. The process of deriving ESCs from an embryo results in the destruction of the embryo. Thus, ESC research necessarily depends upon the destruction of a human embryo.

Despite defendants' attempt to separate the derivation of ESCs from research on the ESCs, the two cannot be separated. Derivation of ESCs from an embryo is an integral step in conducting ESC research. . . . If one step or "piece of research" of an ESC research project results in the destruction of an embryo, the entire project is precluded from receiving federal funding by the Dickey-Wicker Amendment. . . . Accordingly, the Court concludes that, by allowing federal funding of ESC research, the Guidelines are in violation of the Dickey-Wicker Amendment.

(Also see *Sherley v. Sibelius* 2011)

Source: *Sherley v. Sebelius*, Civ. No. 1:09-cv-1575 (RCL). https://ecf.dcd.uscourts.gov/cgi-bin/show_public_doc?2009cv1575-44. Accessed on January 3, 2015.

Sherley v. Sebelius (644 F. 3d 388, 2011)

On April 29, 2011, the U.S. Court of Appeals for the District of Columbia reviewed the decision of Judge Royce Lamberth (2010, the previous document) and reversed his ruling in the case of Sherley v. Sebelius. *The court ruled that:*

We conclude the plaintiffs are unlikely to prevail because Dickey-Wicker is ambiguous and the NIH seems reasonably to have concluded that, although Dickey-Wicker bars funding for the destructive act of deriving an ESC from an embryo, it does not prohibit funding a research project in which an ESC will be used. We therefore vacate the preliminary injunction.

In particular, the court focused on the relative damage that would be caused by the district court injunction on the funding of ESC research:

As we see it, however, a preliminary injunction would in fact upend the status quo. True, the plaintiffs compete with ESC researchers for funding—indeed, that is why they have standing to bring this case, see Sherley I, 610 F.3d at 71–74—but they have been competing with ESC researchers since 2001. The 2009 Guidelines inflict some incremental handicap upon the plaintiffs' ability to compete for NIH money—they point to the additional time and money they must expend and have had to expend since 2001 to meet the additional competition from researchers proposing to use ESCs—but it is necessarily uncertain whether invalidating the Guidelines would result in the plaintiffs getting any more grant money from the NIH. Accordingly, we cannot say that, if the plaintiffs are to litigate this case without the benefit of interim relief, then the 2009 Guidelines will place a significant additional burden upon their ability to secure funding for their research.

The hardship a preliminary injunction would impose upon ESC researchers, by contrast, would be certain and substantial. The injunction entered by the district court would preclude the NIH from funding new ESC projects it has or would have deemed meritorious, thereby inevitably denying other scientists funds they would have received. Even more problematic, the injunction would bar further disbursements to ESC researchers who have already begun multi-year projects in reliance upon a grant from the NIH; their investments in project planning would be a loss, their expenditures for equipment a waste, and their staffs out of a job. The record shows private funding is not generally available for stem cell research but even if, as the district court thought, private donors or investors would provide a reasonable alternative source of funds for ESC researchers, 704 F. Supp. 2d at 72, it remains unclear why such donors or investors would not similarly support the plaintiffs' research using adult stem cells and why the plaintiffs' "very livelihood" instead depends upon obtaining grants from the NIH.

All this is to say the balance of equities s tilts against granting a preliminary injunction. That, combined with our conclusion the plaintiffs have not shown they are likely to succeed on the merits, leads us to hold the district court abused its discretion in awarding preliminary injunctive relief.

Source: *Sherley v. Sebelius,* 2011. No. 10-5287. http://www. cadc.uscourts.gov/internet/opinions.nsf/DF210F382F98EBA C852578810051B18C/$file/10-5287-1305585.pdf. Accessed on January 3, 2015.

Human Cloning Prohibition Act of 2012, H.R. 6623 (2012)

Bills dealing with human cloning have been introduced into the U.S. Congress in 2001, 2003, 2007, and 2012. None has progressed very far in the legislative process. The rationale for the 2012 bill as given in Section 2 of the bill is given here. The prohibition on cloning itself is given in Section 3 and is very similar to that provided in the North Dakota act excerpted earlier.

Section 1

Congress finds that—

(1) some individuals have announced that they will attempt to clone human beings using the technique known as somatic

cell nuclear transfer already used with limited success in sheep and other animals;

(2) nearly all scientists agree that such attempts pose a massive risk of producing children who are stillborn, unhealthy, or severely disabled, and considered opinion is virtually unanimous that such attempts are therefore grossly irresponsible and unethical;

(3) efforts to create human beings by cloning mark a new and decisive step toward turning human reproduction into a manufacturing process in which children are made in laboratories to preordained specifications and, potentially, in multiple copies;

(4) because it is an asexual form of reproduction, cloning confounds the meaning of father and mother and confuses the identity and kinship relations of any cloned child, and thus threatens to weaken existing notions regarding who bears which parental duties and responsibilities for children;

(5) because cloning requires no personal involvement by the person whose genetic material is used, cloning could easily be used to reproduce living or deceased persons without their consent;

(6) creating cloned live-born human children (sometimes called reproductive cloning) necessarily begins by creating cloned human embryos, a process which some also propose as a way to create embryos for research or as sources of cells and tissues for possible treatment of other humans;

(7) the prospect of creating new human life solely to be exploited and destroyed in this way has been condemned on moral grounds by many, including supporters of a right to abortion, as displaying a profound disrespect for life, and recent scientific advances with adult stem cells indicate that there are fruitful and morally unproblematic alternatives to this approach;

(8) in order to be effective, a ban on human cloning must stop the cloning process at the beginning because—

(A) cloning would take place within the privacy of a doctor-patient relationship;

(B) the transfer of embryos to begin a pregnancy is a simple procedure; and

(C) any government effort to prevent the transfer of an existing embryo, or to prevent birth once the transfer has occurred, would raise substantial moral, legal, and practical issues, so that it will be nearly impossible to prevent attempts at reproductive cloning once cloned human embryos are available in the laboratory;

(9) the scientifically and medically useful practices of cloning of DNA fragments, known as molecular cloning, the duplication of somatic cells (or stem cells) in tissue culture, known as cell cloning, and whole-organism or embryo cloning of nonhuman animals are appropriate uses of medical technology;

(10) in the preamble to the 1998 Additional Protocol on the Prohibition of Cloning Human Beings the Council of Europe agreed that the instrumentalisation of human beings through the deliberate creation of genetically identical human beings is contrary to human dignity and thus constitutes a misuse of biology and medicine;

(11) collaborative efforts to perform human cloning are conducted in ways that affect interstate and even international commerce, and the legal status of cloning will have a great impact on how biotechnology companies direct their resources for research and development;

(12) at least 23 countries have banned all human cloning, including Canada, France, and Germany;

(13) the United Nations has passed a declaration calling for all human cloning to be banned by member nations; and

(14) attempts to create cloned human embryos for development of embryonic stem cell lines have been unsuccessful, including the exploitation of over a hundred women in South Korea to provide over 2,000 human eggs without the production of a single stem cell line.

Source: "Text of the Human Cloning Prohibition Act of 2012." Govtrack. US. https://www.govtrack.us/congress/bills/112/hr6623/text. Accessed on July 26, 2014.

Abraham & Veneklasen, et al. v. American Quarter Horse Association (2:12-cv-00103, 2013)

In April 2012, horse breeders Abraham and Veneklasen and two of their companies sued the American Quarter Horse Association (AQHA) in an attempt to force the association to permit cloned horses in their stable to participate in AQHA-approved events. Veneklasen had been producing clones of some of his best horses and wished to enter those clones into AQHA events, but was prohibited from doing so by a regulation adopted by the association in 2004 (then Rule 227(a); later Rule REG106.1). Veneklasen raised two specific antitrust complaints, one that AQHA attempted to restrict trade under the federal Sherman Antitrust Act and one that it attempted to monopolize the market under the Texas Free Enterprise and Antitrust Act. Judge Mary Lou Robinson denied most of the plaintiffs' complaints, but granted the claim of "attempted monopolization." Her reasoning was as follows:

The AQHA asserts a final argument under Section 2: Even if Rule 227(a) is exclusionary and potentially increases or maintains monopoly power, it is not anticompetitive because registration rules are necessary for competition. Without rules, the Quarter Horse would be undefined and the industry would lose coherence. See Hatley, 552F.2d at 654 (holding that rule excluding horses with too much white did not inhibit competition). But at least one court has recognized that where the AQHA stops defining its breed and starts restricting breeding, it can run afoul of antitrust law. Floyd v. Am. Quarter Horse Ass'n, No. 87,589-C, letter ruling at 3 (251st Jud. Dist. of Texas, Dec. 15, 2000). Reproductive limitations do not on their face promote a clearly-defined breed like many physical limitations do. Certainly, a line must be drawn somewhere to distinguish between a traditionally non-white breed and others. Yet where a breed is already physically and genealogically defined, there may be few justifiable reasons to exclude animals that fit these parameters so perfectly that they are indistinguishable from some of the breed's champions.

. . .

Conclusion

The AQHA's motion for summary judgment of the Section 1 conspiracy claim and the Section 2 monopolization claim is DENIED, as it is against any corresponding Texas Free Enterprise and Antitrust Act claims. The AQHA's motion against Plaintiffs' Section 2 attempted monopolization claim is GRANTED, as it is against any corresponding Texas Free Enterprise and Antitrust Act claims.

IT IS SO ORDERED.

Source: *Abraham & Veneklasen, et al. v. American Quarter Horse Association.* http://www.gpo.gov/fdsys/pkg/USCOURTS-txnd-2_12-cv-00103/pdf/USCOURTS-txnd-2_12-cv-00103-0.pdf. Accessed on January 4, 2015. For more details on the case, see Bell, Ryan T. "Clone Case." Bioethicsarchive.georgetown.edu/pcbe/reports/cloningreport/children.html. Accessed on January 4, 2015.

The AQHA appealed this decision of the district court to the Fifth Circuit Court of Appeals, which rendered its judgment on January 14, 2015. In that judgment, the appeals court reversed the decision of the lower court, based on the following logic.

Plaintiffs accuse the AQHA and SBRC of "sham procedures" designed to defeat registration of cloned horses. They refer to a "secret meeting" in January 2012 that, behind the back of AQHA's then president, lay the groundwork for SBRC's official rejection of registration for clones. The only evidence of a meeting in January 2012, however, is an email *from* the president inviting certain SBRC members to an official meeting of AQHA's Executive Committee meeting to discuss cloning. There was nothing secret about it. Even more telling, there is no testimony about what transpired at the not-so-secret meeting.

Plaintiffs contend that AQHA "stacked" the SBRC with hand-selected industry leaders with interests in conflict with cloning. As has been noted, the committee was never shown to have a voting majority of members with interests in elite Quarter Horses, although most of its members, unsurprisingly, have been breeders of Quarter Horses. In any event, Plaintiffs failed to explain why the selection of SBRC members was not as consistent with permissible activity as it was with impermissible activity; selecting industry leaders who are knowledgeable about breeding for a committee focused on registration of the breed seems quite reasonable. . . . Plaintiffs did not produce evidence tending to exclude the possibility of a decision arrived at by independent, not illegally concerted action.

Finally, the plaintiffs focus on a "plan" to delay and ultimately reject cloned horse registration that allegedly appeared in the handwritten notes of AQHA's executive director Don Treadway. The eight pages of random, scrawled notes span nearly two years and derive from various meetings and conversations. While they reveal Treadway's thinking and concerns others expressed about cloning and AQHA's possible reaction to it, they contain no "smoking gun" referencing any agreement within AQHA or its membership to restrain the market for elite Quarter Horses.

Reasonable jurors, in sum, could not draw any inference of conspiracy from the evidence presented, because it neither tends to exclude the possibility of independent action nor does it suggest the existence of any conspiracy at all. In the absence of substantial evidence on the issue of an illegal conspiracy to restrain trade, AQHA's JMOL motion should have been granted.

. . .

For these reasons, we **REVERSE** and **RENDER** judgment for the Appellant.

Source: Abraham & Veneklasen Joint Venture; Abraham Equine, Incorporated; Jason Abraham, Plaintiffs—Appellees v. American Quarter Horse Association. No. 13–11043. http://www.ca5.uscourts.gov/opinions%5Cpub%5C13/13-11043 -CV0.pdf. Accessed on January 18, 2015.

Stem Cell Research Advancement Act of 2013, H.R. 2433 (2013)

For more than a decade, a variety of bills on cloning and stem cell research have been introduced into the U.S. Congress, with a variety of results. None, however, has ever been enacted into law. The most recent attempt to provide federal funding for stem cell research has been the Stem Cell Research Advancement Act, which was introduced in Congress in 2011 and 2013. In both cases, the act was relatively simple in language and appeared to be a pro forma effort, given the low probability that it would ever be enacted. The core of the bill in both cases was as follows:

A Bill

To amend the Public Health Service Act to provide for human stem cell research, including human embryonic stem cell research, and for other purposes.

1. Short title
    This Act may be cited as the Stem Cell Research Advancement Act of 2013.
2. Findings
    The Congress finds as follows:

(1) On March 9, 2009, President Barack Obama issued Executive Order 13505, entitled Removing Barriers to Responsible Scientific Research Involving Human Stem Cells.

(2) On July 7, 2009, the National Institutes of Health issued guidelines on human stem cell research.

(3) The scientific field of stem cell research is continually advancing.

3. Human stem cell research

Part H of title IV of the Public Health Service Act (42 U.S.C. 289 et seq.) is amended by inserting after section 498D the following:

498E. Human stem cell research

(a) In general

Notwithstanding any other provision of law, the Secretary shall conduct and support research that utilizes human stem cells, including human embryonic stem cells.

(b) Eligibility

To be eligible for use in research under subsection (a), human embryonic stem cells must have been derived from cells from human embryos that—

(1) were created using in vitro fertilization for reproductive purposes and are no longer needed for those purposes; and

(2) were donated by the individuals who sought reproductive treatment with written and voluntary informed consent for the embryos to be used for research purposes and without receiving any financial or other inducements to make the donation.

Source: Text of the Stem Cell Research Advancement Act of 2013. GovTrack.us. https://www.govtrack.us/congress/bills/113/hr2433/text. Accessed on January 2, 2015.

Research

Introduction

Researchers have been writing about cloning for centuries. Only in the last half century, however, has the topic become one of common interests among the general public. Intelligent discussions of cloning technology and the issues related to it, however, require that individuals be fully informed as to what cloning involves, how cloning is done, what the potential risks associated with cloning may be, what applications are (and ultimately will be) available for cloning, and what the risks and benefits of cloning can be for the world. This chapter provides an annotated bibliography of a sample of the books, reports, articles, and Internet sources that deal with the topic of cloning. A number of references occur in more than one format, as articles and as Web pages, for example. In such instances, both references for the item are provided here.

Books

Alexandrov, Kirill, and Wayne A. Johnson, Eds. 2014. *Cell-Free Protein Synthesis: Methods and Protocols*. New York: Humana Press; Springer.

A coyote cloned by South Korean stem cell scientist Hwang Woo-suk and his team is pictured on a farm at a wildlife protection center in Pyeongtaek, south of Seoul. Hwang donated eight coyotes that he and his research team at the Sooam Biotech Research Foundation cloned to help the endangered species, to the Gyeonggi provincial government. (Jo Yong-Hak/Corbis)

This book is a technical summary of the range of research taking place for the production of protein products using laboratory technologies based on cloning.

Anthes, Emily. 2014. *Frankenstein's Cat: Cuddling up to Biotech's Brave New Beasts*. New York: Scientific American/Farrar, Straus and Giroux, 2014.

The author provides a highly readable account of the many types of animals that have been cloned thus far, along with a discussion as to the utility and value of such research.

Banerjee, Ena Ray. 2014. *Perspectives in Regenerative Medicine*. New Delhi: Springer.

This book is designed as a textbook for undergraduate students in the life sciences. It provides a broad, general introduction to the subject of stem cell research and regenerative medicine with chapters on topics such as the stem cell and its niche, study models for stem cells and the effects of trace elements, the use of stem cells in drug screening, differentiation of embryonic stem cells, and model organisms in science and research.

Barber, Nicola. 2013. *Cloning and Genetic Engineering*. New York: Rosen Publishing's Rosen Central.

This book provides a good general introduction for young adults to the topic of cloning and genetic engineering.

Bradley, James T. 2013. *Brutes or Angels: Human Possibility in the Age of Biotechnology*. Tuscaloosa: University of Alabama Press.

The author begins this book with a general introduction to basic topics in biology, such as the nature of cells and molecules, and then moves on to a review of the developments in modern biology that now allow researchers to make fundamental changes in the nature of human life. He then outlines some future possibilities for the nature of human life and discusses the ethical, social, and other consequences of such changes.

Brown, T.A. *Gene Cloning and DNA Analysis*, 6th ed. New York: Wiley-Blackwell.

> This book provides a technical introduction to the topic of gene cloning and DNA analysis, with three major sections, the first of which provides an introduction to the technology of gene cloning and DNA analysis. Sections two and three, then, discuss the application of gene cloning and DNA analysis to research and biotechnology, respectively.

Calegari, Frederico, and Claudia Waskow, eds. 2014. *Stem Cells: From Basic Research to Therapy*. Boca Raton, FL: CRC Press, Taylor & Francis Group.

> This book provides a collection of papers dealing with technical aspects of stem cell research, divided into three major parts: basic stem cell biology, tissue formation during development, and model organisms.

Cibelli, Jose, et al., eds. 2014. *Principles of Cloning*, 2nd ed. Amsterdam: Elsevier/Academic Press.

> This is the second edition of a popular and highly regarded text that provides a general introduction to the topic of cloning. Individual chapters deal with topics such as the history of cloning, the methods of cloning, cloning of various species, applications, complementary technologies, somatic cell nuclear transfer, and ethical and legal issues associated with cloning.

Firth, Lisa, ed. 2013. *Biotechnology and Cloning*. Cambridge, UK: Independence Educational Publishers.

> This book is volume 211 in the publisher's "Issues: See the Larger Picture" series. Its three chapters deal with the fundamentals of biotechnology, animal cloning, and human cloning, discussing the technical aspects of each, along with social and ethical issues related to each subject.

Fletcher, Amy Lynn. 2014. *Mendel's Ark: Biotechnology and the Future of Extinction*. New York: Springer.

The author begins by discussing the issues involved in the extinction of plant and animal life overall worldwide today. She then goes on to review some ways in which the threat of extinction can be reduced for some species, with particularly strong chapters on the cloning of animals that have long been extinct and animals that have only recently become extinct or are now threatened with extinction.

Friese, Carrie. 2013. *Cloning Wild Life: Zoos, Captivity, and the Future of Endangered Animals.* New York: New York University Press.
Friese provides a brief introduction to the topic of cloning in general and then focuses on the application of cloning to extinct and endangered species, especially those held by zoos and other captive breeding sites, with special attention to the unique issues that such sites must deal with in the cloning of animals.

Gerdes, Louise. 2014. *Human Genetics.* Farmington Hills, MI: Greenhaven Press.
This book is written for young adults and is part of the publisher's Opposing Viewpoints series. Books in this series provide a variety of perspective essays on important scientific and social topics, such as, in this case, stem cell research and cloning.

Hogan, Kelly A. 2009. *Stem Cells and Cloning,* 2nd ed. San Francisco: Pearson/Benjamin Cummings.
This short book provides a general introduction to the topic of stem cells and cloning, possible applications of these technologies, ethical issues involved in such applications, and a look at the future of research in both fields.

Hogle, Linda F., ed. 2014. *Regenerative Medicine Ethics: Governing Research and Knowledge Practices.* New York: Springer.
This collection of essays examines a wide variety of ethical issues relating to the practice of regenerative medicine, such as collaborative practices, patenting, repositories for

data, protection of human participants, and early-stage research. Two appendices provide very useful compilations of international policies on stem cell research and use and state personhood laws.

Hug, Kristina, and Göran Hermerén, eds. *Translational Stem Cell Research: Issues beyond the Debate on the Moral Status of the Human Embryo.* New York: Humana Press, 2011.

The philosophy behind this book is that stem cell research has long been embroiled in moral and ethical issues over the use of human embryonic stem cells and that it is now time to go beyond these debates to take a broader view of the topic, with special focus on the progress in stem cell research that is taking place both in the laboratory and in practice. The 10 sections of the book deal with topics such as what is possible today and what still is to be achieved, translating stem cell research from bench to bedside, creation of human-animal entities for research, stem cell banking, the funding of stem cell research, patenting of stem cell results, communication to the general public about stem cell research, psychological implications of stem cell research, ethical implications, and the future of stem cell research.

Humber, James M., and Robert F. Almeder, eds. 1998. *Human Cloning.* Totowa, NJ: Humana Press.

This somewhat dated book is, nonetheless, an excellent resource on the social, ethical, philosophical, and theological issues related to the reproductive cloning of humans. The six scholarly papers that make up this special issue of the journal *Biomedical Ethics Review* provide a good historical background on the ethical issues relating to human reproductive cloning, along with more detailed discussions of the ethical and philosophical issues related to the practice.

Hyun, Insoo. 2013. *Bioethics and the Future of Stem Cell Research.* New York: Cambridge University Press.

The author notes that social, ethical, political, and other questions have long troubled research on embryonic stem cell research. He also points out that in spite of these problems, stem cell research has continued at a significant pace. He recommends that interested observers broaden the scope of their questions about the ethical implications of stem cell research to include topics such as research using adult induced pluripotent stem cells, the issues posed by clinical trials, and stem cell "tourism." The book provides an interesting multidisciplinary look at stem cell research from the standpoint not only of science, technology, and ethics but also of the history of science, literature, and philosophy.

Jensen, Eric Allen. 2014. *The Therapeutic Cloning Debate: Global Science and Journalism in the Public Sphere.* Farnham, UK: Ashgate.

This book focuses on the role played by the media in the debate over the use of cloning for therapeutic applications. It provides a nice background history on the technology of therapeutic cloning, along with a review of the way this technology has been described and discussed in the popular press. It then analyzes a number of factors related to the way that the press presents the issues surrounding therapeutic cloning.

Knoepfler, Paul. 2013. Stem *Cells: An Insider's Guide.* Singapore: World Scientific.

This book is written for the reader with little or no background in cloning, stem cell research, or biology in general. It provides a clear and enjoyable introduction to the field of stem cell research along with virtually every related topic that can be imagined, such as types of stem cells, the use of stem cells in dealing with the process of ageing, laws and regulations concerning stem cell research, ethical issues related to the use of stem cells, a proposed patient's bill of rights with regard to the use of stem cells, and guidance for possible stem cell therapies.

Kumar, Lephen Thank. 2014. *A Christian Response to Human Cloning: Edenic Eve versus Clonaid Eve.* New Delhi: Christian World Imprints.

> This book attempts to present an overview of the current debate over human reproductive cloning "from a Christian standpoint." It discusses the issue by comparing the Biblical Eve, whom God is said to have created at Earth's creation, to the baby purported to have been created by the Clonaid Company in 2002.

Laimer, Margit, and Waltraud Rücker, eds. 2003. *Plant Tissue Culture: 100 Years since Gottlieb Haberlandt.* Vienna: Springer.

> This book provides an excellent review of the earliest days of research using plant tissue cultures, including not only commentaries on some of the most important figures in the field but also reprints of a number of the most important papers of the time.

Levine, Aaron D. 2012. *Cloning.* New York: Oneworld Publications.

> This book is one in the publisher's series A Beginner's Guide, which provides a general introduction to a wide variety of current social issues. Various chapters in the book discuss the technology of cloning and stem cell research, animal cloning and its developments, ethical issues related to cloning, and the possible future of cloning.

Macintosh, Kerry Lynn. 2012. *Human Cloning: Four Fallacies and Their Legal Consequences.* Cambridge: Cambridge University Press.

> The author posits the argument that the arguments offered in opposition to the cloning of humans is based more on intuition than on scientific facts and that the conclusions that people have drawn based on these intuitive beliefs have been counterproductive for individuals on all sides of the political spectrum. She claims that human cloning can actually be a useful and valuable technology for the future.

Mummery, Christine, et al. 2014. *Stem Cells: Scientific Facts and Fiction*, 2nd ed. London: Elsevier, Academic Press.

> This book provides an excellent introduction to the topic of stem cells and cloning for the nonexpert in the field. The authors provide a good introduction to the relevant biology, including discussions of the cell, biochemistry, reproduction, and development, before introducing the topic of stem cells, the ways in which they are used in research, and some of the many applications they do have and may have in the future as a tool in regenerative medicine.

Oksanen, Markku, and Helena Siipi, eds. 2014. *The Ethics of Animal Re-creation and Modification: Reviving, Rewilding, Restoring*. New York: Palgrave Macmillan.

> This anthology contains a group of papers dealing with current interests in restoring certain extinct species using the technologies available from cloning. The papers raise issues about changing the focus of conservation efforts from protection of species to restoration and how this will affect current efforts in the former direction.

Park, Alice. 2012. *The Stem Cell Hope: How Stem Cell Medicine Can Change Our Lives*. New York: Hudson Street Press.

> This book provides an excellent general introduction to the topic of stem cell research and regenerative medicine for the average reader. It contains chapters on the history of cloning, political factors involved in stem cell research, fraudulent research that has been conducted in the field, and the promise that stem cell therapies may have for enhanced human health in the future.

Peters, Ted, Karen Lebacqz, and Gaymon Bennett. 2008. *Sacred Cells?: Why Christians Should Support Stem Cell Research*. Lanham, MD: Rowman & Littlefield Publishers.

> These three faculty members at the Center for Theology and the Natural Sciences at the Graduate Theological

Union in Berkeley, California, say that opposition by Christians to stem cell research is based on some major understandings of theology, stem cell research, and the relationship between the two. They suggest that once these misunderstandings are cleared up, it will become clear why Christians should support stem cell research.

Reed, Don C. 2015. *Stem Cell Battles Prop 71 and Beyond: The Struggle to Preserve and Advance Stem Cell Research: How the Fight against Chronic Disease and Disability Can Be Won*. Singapore: World Scientific Publishing Company.

Reed makes use of the battle over California Proposition 71 to discuss the political debate over stem cell research and reviews some of the promises that such research can hold for improvements in human healthcare in the future.

Smith, Roberta H. 2013. *Plant Tissue Culture: Techniques and Experiments*, 3rd ed. London: Academic Press.

This popular and excellent book provides a comprehensive introduction to the topic of plant tissue culture, including a very interesting and useful introductory chapter on the history of this field of research.

Thomas, Isabel. 2013. *Should Scientists Pursue Cloning?* London: Raintree.

This book, intended for juvenile readers, provides an overview of the ethical issues posed by the possibility of modern cloning technology.

Turksen, Kursad, ed. 2014. *Adult Stem Cells*, 2nd ed. New York: Humana Press.

This technical book provides detailed discussions of the extraction and use of adult stem cells from a number of body systems and possible therapeutic use of such cells.

Valla, Svein, and Rahmi Lale, eds. 2014. *DNA Cloning and Assembly Methods*. New York: Humana Press.

This technical volume offers a view into the wide range of practices that professionals in the field now routinely use in their own research.

Van der Wolf, Willem Jan, and Relinde Van Laar, eds. 2008, 2011. *Cloning and Stem Cell Research: Legal Documents*. Oisterwijk, The Netherlands: Wolf Legal Publishers.
This title is available in six volumes dealing with legislative developments and actions related to stem cell research. Although now somewhat out of date, these books are almost certainly one of the most complete and valuable resources on the legal status of stem cell research and cloning currently available.

Wilmut, Ian, and Roger Highfield. 2006. *After Dolly: The Uses and Misuses of Human Cloning*. New York: W. W. Norton.
Wilmut, writing with science journalist Highfield, tells an interesting story of the cloning of the first sheep, Dolly, along with a number of related topics, such as an overview of the history of cloning, some consequences resulting from the announcement of this achievement, some ethical issues related to cloning, the wisdom (or not) of human reproductive cloning, and possible future directions for cloning research.

Woestendiek, John. 2010. *Dog, Inc.: The Uncanny inside Story of Cloning Man's Best Friend*. New York: Avery.
The author reviews the history of efforts to clone domestic and farm animals, beginning with the attempted cloning of the dog Missy by multibillionaire John Sperling and his wife Joan Hawthorne. He continues the story with episodes from the attempted (and often successful) cloning of the first cloned cat and sheep (Dolly) with discussions of the technical, social, and ethical issues involved in this line of research.

Articles

Blum, Hubert E. 2014. "Cell Therapies and Regenerative Medicine." *Hepatology International*. 8(2): 158–165.

This review article provides the general background for the development of a number of cell technologies that have made possible a huge leap forward in the field of regenerative medicine. The author outlines some of the ways in which these advances may have medical applications in the near future.

Briggs, Robert, and Thomas J. King. 1952. "Transplantation of Living Nuclei from Blastula Cells into Enucleated Frogs' Eggs." *Proceedings of the National Academy of Science of the United States*. 38(5): 455–463. Also available at http://www.pnas.org/content/38/5/455.full?ijkey=ce224839890f10a7dc248153ae51007044200a1f&keytype2=tf_ipsecsha. Accessed on July 27, 2014.

This article is one of the most famous in the history of cloning. It describes the transfer of nuclei from the northern leopard frog, *Rana pipiens*, into enucleated cells, a process that results in the production of embryos and then adult frogs identical to those from which the nuclei were taken.

Burchell, Melissa S. 2004. "America's Struggle to Develop a Consistent Legal Approach to Controversial Human Embryonic Stem Cell Research and Therapeutic Cloning: Are the Politics getting in the Way of Hope." *Syracuse Journal of International Law and Commerce*. 32(1): 133–161.

The author reviews the recent history of therapeutic cloning and the ethical issues involved in its use. She then compares the legal and policy approaches taken by the U.S. and British governments for dealing with the technology. She suggests, in conclusion, that the United States "adopt a legal scheme similar to that in Britain regarding human embryonic stem cell research and therapeutic cloning."

Chan, Albert Wai-Kit, Alice Yuen-Ting Wong, and Hon-Man Lee. 2014. "A Patent Perspective on US Stem Cell Research." *Nature Biotechnology*. 32(7): 633–637.

In this article, the authors review legal decisions that have impact on the practice of stem cell research and application, including cases decided by the U.S. Supreme Court, as well as policy decisions made by various branches of the U.S. government.

Cohen, Shlomo. 2014. "The Ethics of De-Extinction." *Nanoethics*. 8(2): 165–178.

Cohen notes the rapid progress being made in the field of cloning and suggests that deciding on the ethical implications of cloning extinct species is a complex ethical issue that has at least five possible dimensions. He suggests that the question of cloning extinct species "repeatedly tests the limits of our ethical notions."

Cyranoski, David. 2014. "Cloning Comeback." *Nature*. 505 (7484): 468–471. http://www.nature.com/news/cloning-come back-1.14504. Accessed on January 18, 2015.

Cyranoski reviews the troubled history of South Korean researcher Hwang Woo-suk who fraudulently reported successful human cloning experiments in the early 2000s, but has gone on to conduct highly successful work in the field of the cloning of nonhuman primates.

DeSouza, Natalie. 2013. "Single-cell Genetics." *Nature Methods*. 10(9): 820.

This article describes a promising new approach to stem cell research that has since become increasingly popular. The original paper to which this article refers is by Wills et al., listed later.

Dröscher, Ariane. 2014. "Images of Cell Trees, Cell Lines, and Cell Fates: The Legacy of Ernst Haeckel and August Weismann in Stem Cell Research." *History and Philosophy of the Life Sciences*. 36(2): 157–186.

This essay explores some of the earliest thoughts and writings about the concept of stem cells in the work of Ernest Haeckel and August Weismann.

The Ethics Committee of the American Society for Reproductive Medicine. 2012. "Human Somatic Cell Nuclear Transfer and Cloning." *Fertility and Sterility*. 98(4): 804–807.

This document is the report of a special committee on the ethics of cloning in human reproductive technologies. The report concludes that the use of cloning is unethical for infertility treatment "due to concerns about safety; the unknown impact of SCNT on children, families, and society; and the availability of other ethically acceptable means of assisted reproduction."

Fadel, Hossam E. 2012. "Developments in Stem Cell Research and Therapeutic Cloning: Islamic Ethical Positions, a Review." *Bioethics*. 26(3): 128–135.

The author reviews attitudes toward cloning and stem cell research in the Islamic world and briefly compares those attitudes toward those in non-Islamic countries. He points out in general that Islamic scholars approve of therapeutic cloning but points out the conditions under which such research may legitimately be conducted in the Islamic world.

Foley, Elizabeth Price. 2000. "The Constitutional Implications of Human Cloning." *Arizona Law Review*. 42(3): 647–730.

The author provides a very detailed legal analysis of the challenges posed by the technical possibilities of human reproductive cloning. She concludes that "the current legal regime appears prepared to handle human cloning. Moreover, should Americans wish to ban the practice, they may not be able to do so consistent with the current Constitution."

Foong, Patrick. 2012. "The Need to Pay Egg Donors for Use in Therapeutic Cloning/Somatic Cell Nuclear Transfer (SCNT) Research." *Legal Issues in Business*. 14: 3–8.

Australia's Research Involving Human Embryos Act of 2002 was amended in 2006 to permit the use of human eggs in therapeutic cloning research. But the act prohibited researchers from paying donors for such eggs. Foong presents arguments in this article as to why that policy should be reversed and payment should be permitted for the purchase of human eggs.

Gautheret, Roger J. 1983. "Plant Tissue Culture: A History." *The Botanical Magazine.* 96(4): 393–410.

This article offers a personalized history of plant tissue culture research by one who was a major player in that field for many years.

Gonçalves, N. N., C. E. Ambrósio, and J. A. Piedrahita. 2014. "Stem Cells and Regenerative Medicine in Domestic and Companion Animals: A Multispecies Perspective." *Reproduction in Domestic Animals.* 49: 2–10.

The authors point out some of the problems in using mice and other small animals in stem cell research because such animals may differ significantly from humans who are the ultimate targets of such research. They point out that domestic animals may be closer in characteristics to humans than are mice and that protocols for using such animals in stem cell research should and could be developed.

Harris, John. 2014. "Time to Exorcise the Cloning Demon." *Cambridge Quarterly of Healthcare Ethics.* 23(1): 53–62.

The author notes that a breakthrough in methods for producing stem cells produced, among other responses, a new flood of alarms about the possible use of cloning to reproduce human beings. He says that such "inflammatory suggestion[s]" are largely unnecessary, because human reproductive cloning is already illegal in most countries of the world, and they only serve to interrupt or interfere with the more appropriate use of cloning for therapeutic purposes.

Hoppe, Philipp S., Daniel L. Coutu, and Timm Schroeder. 2014. "Single-cell Technologies Sharpen up Mammalian Stem Cell Research." *Nature Cell Biology.* 16(10): 919–927.

This article provides an excellent example of the way in which a new technology—single-cell technology—has improved the results of stem cell research. For further background on the technology, also see Wills, below, and de Souza, listed earlier.

Hua, Song, Henry Chung, and Kuldip Sidhu. 2014. "Human Therapeutic Cloning, Pitfalls and Lack Luster Because of Rapid Developments in Induced Pluripotent Stem Cell Technology." *Asian Biomedicine.* 8(1): 5–10.

> The authors provide a good review of both somatic cell nuclear transplantation (SCNT) and the use of induced pluripotent stem cells (iPSCs) for the purpose of producing clones. Then they compare the benefits and disadvantages of using each type of technology and discuss the circumstances in which each is more appropriate.

Ikemoto, Lisa C. 2014. "Can Human Embryonic Stem Cell Research Escape Its Troubled History?" *The Hastings Center Report.* 44(6): 7–8.

> The author reviews the way in which social, ethical, religious, and other concerns have tended to limit the progress of stem cell research and asks if new discoveries in which adult cells have been used to produce human embryonic stem cells will reverse that pattern.

Isasi, R., et al. 2014. "Identifiability and Privacy in Pluripotent Stem Cell Research." *Cell Stem Cell.* 14(4): 427–430.

> Data sharing is a common and expected aspect of most scientific research. Yet, some recent developments in stem cell research have raised new issues with regard to how, when, and to what extent data should or can be shared among researchers. The authors of this article proposed some new guidelines to determine the best policies for data sharing in stem cell research.

Jensen, Eric. 2012. "Scientific Sensationalism in American and British Press Coverage of Therapeutic Cloning." *Journalism and Mass Communication Quarterly.* 89(1): 40–54.

> Articles about cloning often have a tendency to include an element of emotionalism, "hype," sensationalism, or other nonscientific and nontechnical reporting. The author of this article reviewed 5,128 news articles in the

British and American press to determine the extent to which those articles included some of these elements.

Kim, M. J., et al. 2012. "Lessons Learned from Cloning Dogs." *Reproduction in Domestic Animals*. 47(Suppl 4): 115–119.

The authors review the results of their successful cloning of a number of domestic dogs. They report that the cloned dogs are like noncloned cousins in nearly all respects, and the technology that has been developed for this research shows great promise for the purpose of conserving endangered species, treating sterile canids or aged dogs, improving reproductive performance of valuable individuals, and generating disease model animals.

Krause, Kenneth W. 2012. "New Life for Human (Therapeutic) Cloning? Whence Come the Eggs?" *Skeptical Inquirer*. 36(2): 28–30.

This article describes the discovery of the process by which induced pluripotent stem cells (iPSCs) can be produced and how this discovery may have significantly changed the future for therapeutic cloning. He also discusses the relative advantages and disadvantages of using iPSC and SCNT procedures in cloning research and applications.

Lévesque, Maroussia, et al. 2014. "Stem Cell Research Funding Policies and Dynamic Innovation: A Survey of Open Access and Commercialization Requirements." *Stem Cell Reviews and Reports*. 10(4): 455–471.

The pressure for finding and making use of specific applications of stem cell research in medical applications has raised issues about the relevance and implementation of classical demands for data sharing among researchers in the field. The authors of this article survey the way in which the results of stem cell research are being implemented in various countries of the world and identify a number of factors that are inhibiting the normal elements of scientific research. They suggest that regulatory agencies may need to develop new policies that ensure the free flow of information in the field of stem cell research.

Mathur, Shivani, et al. 2014. "Stem Cell Research: Applicability in Dentistry." *The International Journal of Oral & Maxillofacial Implants*. 29(2): e210–e219.

> Although one may tend to think of the applications of stem cell research in medicine, the field also has some obvious and important applications in the field of dentistry, especially in the regeneration of damaged periodontal tissue, bone, pulp, and dentin. This article provides an excellent review of progress in the field of stem cell therapy in dental applications.

McKinnell, Robert G., and Marie A. DiBeradino. 1999. "The Biology of Cloning: History and Rationale." *BioScience*. 49(11): 875–885.

> Although now very dated, this article provides an excellent review of the early history of cloning up to about 1999, along with an explanation of the scientific value and possible applications of the technology.

Navratyil, Zoltán. 2013. "Legal and Ethical Aspects of Human Reproductive Cloning." *Acta Juridica Hungarica*. 54(1): 106–117.

> This very interesting article attacks an issue that is seldom discussed as clearly as it is here: the nearly universal ban on human reproductive cloning. The author reviews laws (or the lack of laws) on human reproductive cloning in Germany, Great Britain, Hungary, and the United States and comes to the conclusion that "the statutory prohibition of reproductive cloning often does not correspond to the biological facts." That failure, he says, may well lead to legal and policy complications and confusion. He concludes the article with an analysis for the reasons that human reproductive cloning is so strongly censured almost everywhere.

Niemann, H., and A. Lucas-Hahn. 2012. "Somatic Cell Nuclear Transfer Cloning: Practical Applications and Current Legislation." *Reproduction in Domestic Animals.* 47(Suppl. s5): 2–10.

>The authors briefly review the process of somatic cell nuclear transplantation in domestic animals and then talk in some detail about the many practical applications of this technology for this group of animals. They then outline in some detail the legislative actions that have been taken, primarily in Europe, to deal with the new problems created by the cloning of animals for the production of food.

Owen-Smith, Jason, Christopher Thomas Scott, and Jennifer B. McCormick. 2012. "Expand and Regularize Federal Funding for Human Pluripotent Stem Cell Research." *Journal of Policy Analysis and Management.* 31(3): 714–722.

>The authors provide a brief history of the discovery and development of research on embryonic human stem cells, emphasizing the potential therapeutic value they may have for a variety of human diseases and disorders. They then suggest that federal funding for stem cell research should be expanded and better regulated for three reasons. First the application of stem cell therapy to real-life medical conditions is finally beginning to appear, and enhanced legislation can only accelerate the pace of such developments. Second, the ethical basis for stem cell research now seems to be largely settled and widely accepted, and expanded federal research can only codify this ethical philosophy. Finally, funding uncertainties may create even more serious problems for researchers than concerns over or lack of regulation about stem cell research.

Poon, Peter N. 2000. "Evolution of the Clonal Man: Inventing Science Unfiction." *Journal of Medical Humanities.* 21(3): 159–173.

>This fascinating article considers the way in which humans have thought about the possibilities of making exact

copies of themselves—clones—throughout history. He takes note of the fact that such imaginings were purely science fiction until the cloning of the sheep Dolly, at which point humans were faced with far more realistic technical, social, and ethical considerations of what human cloning might actually be like.

Ratajczak, Mariusz Z., et al. 2014. "New Advances in Stem Cell Research: Practical Implications for Regenerative Medicine." *Polskie Archiwum Medycyny Wewnetrznej*. 124(7–8): 417–426.

This article discusses some of the possible sources of stem cells for use in research, a review of the mechanisms by which stem cells may produce their therapeutic effects and the results of some of the first clinical trials with stem cells.

Rhodes, Suzanne H. 2003. "The Difficulty of Regulating Reproductive and Therapeutic Cloning: Can the United States Learn Anything from the Laws of Other Countries." *Penn State International Law Review*. 21(2): 341–361.

Rhodes explains what the legal and policy issues are with respect to both therapeutic and reproductive cloning and suggests how and why it might be helpful for the United States to consider the approaches taken by some other countries around the world in dealing with this issue.

Rodriguez, Ramon M., Pablo J. Ross, and Jose B. Cibelli. 2012. "Therapeutic Cloning and Cellular Reprogramming." *Advances in Experimental Medicine and Biology*. 741: 276–289.

The authors describe the technological procedures that are used by which cells can be reprogrammed to produce genetically identical cells for use in regenerative medical procedures.

Rosen, Michael R., et al. 2014. "Translating Stem Cell Research to Cardiac Disease Therapies: Pitfalls and Prospects for Improvement." *Journal of the American College of Cardiology*. 64(9): 922–937.

This extended article is a collection of papers about progress in the use of stem cells for the treatment of cardiac diseases. The first paper is a review of technical developments in the research and application of the use of stem cells for this purpose. The second and third papers provide a number of reactions from experts in the field on the first paper. The final paper discusses possible future directions for research in this field.

Sandler, Ronald. 2014. "The Ethics of Reviving Long Extinct Species." *Conservation Biology*. 28(2): 354–360.

The author begins this article by reviewing a number of arguments both in favor of and in opposition to the cloning of animals that have been extinct for very long periods of time, even assuming the possibility that such research can be successfully completed. He concludes that the ethical legitimacy of carrying out such research varies from case to case, and the conditions that obtain in each possible situation should be weighed in the balance in deciding whether or not to carry out such a project.

Shapshay, Sandra. 2012. "Procreative Liberty, Enhancement and Commodification in the Human Cloning Debate." *Health Care Analysis: An International Journal of Health Care Philosophy and Policy*. 20(4): 356–366.

The author takes note of the vigorous debate in U.S. society with regard to human reproductive cloning and suggests that those who are opposed to and those who favor the practice may actually have a good deal more in common than they realize or are willing to admit. She points out how recognizing these common positions could advance the debate over human reproductive cloning.

Shikai, Yuriko Mary. 2004. "Don't Be Swept Away by Mass Hysteria: The Benefits of Human Reproductive Cloning and Its Future." *Southwestern University Law Review*. 33(Part 2): 259–284.

The author presents some strong arguments that human cloning should not be prohibited and, in fact, that there are some very good reasons with going ahead with research in the field and its ultimate implementation as an accepted medical technology.

Siegel Bernard, and Arnold I. Friede. 2013. "The U.S. Food and Drug Administration Should Solidify the Legal Basis for Its Authority over Reproductive Cloning." *Stem Cells and Development*. 22(Suppl. 1): 46–49.

The authors take note of the fact that there are still no federal laws in the United States about reproductive or therapeutic cloning. As progress in these fields continue, they say, efforts should be made to clarify U.S. policy on both procedures. In particular, they recommend that the U.S. Food and Drug Administration "authoritatively establish its jurisdiction over human reproductive cloning so as to foster the life-saving potential of therapeutic cloning."

Smith, Mark H. 2014. "Stem Cell Research Review." *Stem Cell*. 5(1): 77–91.

The author provides a very detailed and comprehensive review of the fundamental information regarding stem cell research. Of special value is the bibliography of more than 160 relevant articles, book chapters, and other references.

Svendsen, Clive N. 2013. "Back to the Future: How Human Induced Pluripotent Stem Cells Will Transform Regenerative Medicine." *Human Molecular Genetics*. 22(R1): R32–R38.

The author reviews the history of induced pluripotent stem cells and explains how the new technology opens a whole new area of research for regenerative medicine.

Thorpe, Trevor. 2007. "History of Plant Tissue Culture." *Molecular Biotechnology*. 37(2): 169–180.

This article is one of the most complete accounts of the history of plant tissue culture research, dating back to the work of Gottlieb Haberlandt in the early 20th century.

Wills, Quin F., et al. 2013. "Single-Cell Gene Expression Analysis Reveals Genetic Associations Masked in Whole-Tissue Experiments." *Nature Biotechnology.* 31(8): 748–752.

This article describes a new technology for use in stem cell research that involves the use of single cells rather than complete tissues or whole organisms. For an overview of the article and the line of research, see also de Souza, listed earlier. For an example of the use of single-cell technology, see also the article by Hoppe, Coutu, and Schroeder, listed earlier.

Wilmut, Ian. 2010. "Cloning." *New Scientist.* 207(2772): i–viii.

Wilmut, the researchers who directed the cloning of the sheep Dolly, provides an excellent overview of the field of cloning, including a good review of the history of the technology that considers the contributions of Hans Driesch, Steen Willadsen, Hans Spemann, and some of Wilmut's colleagues. The article also contains some interesting speculations on the possible future of cloning.

Reports

Various departments of the U.S. and other governments have produced a number of reports on cloning and stem cell research, only a few of which are listed here. For more information on such reports, see the Web sites of the agencies involved, especially the Congressional Research Service (http://www.loc.gov/crsinfo/research/).

"Animal Cloning: Reasoned Opinion." 2014. House of Commons [Great Britain]. European Scrutiny Committee. London: Stationery Office.

This report sets out the British government's official policy about the cloning of animals in the country and regulations related to the export and import of such animals into and out of the country.

Berger, Adam C., Sarah H. Beachy, et al. 2014. "Stem Cell Therapies: Opportunities for Ensuring the Quality and Safety of Clinical Offerings: Summary of a Joint Workshop." Washington, DC: National Academies Press.

> This book summarizes the results of a workshop on the current status of stem cell research in the United States sponsored by the Board on Health Sciences Policy of the Institute of Medicine, Board on Life Sciences of the National Research Council, and International Society for Stem Cell Research. The primary objectives of the meeting were to discuss the current use of unregulated stem cell therapies, examine the type of stem cell therapies currently in use, assess the risks and benefits of such therapies, review the evidence needed to justify the use of stem cell therapies, evaluate legal issues related to the use of stem cell therapies, and discuss ways of ensuring that patients receive high-quality stem cell treatments.

Duffy, Diane T. 2002. "Background and Legal Issues Related to Stem Cell Research." Congressional Research Service. Washington, DC: Congressional Research Service.

> This report was prepared primarily in response to the decision by President George W. Bush to allow the use of federal funds for certain types of stem cell research. The report provides a brief history and review of the current status of stem cell research in the United States. The report is available online at http://www.law.umaryland.edu/marshall/crsreports/crsdocuments/RS21044.pdf. Accessed on January 20, 2015.

Genetics Policy Institute. Year varies. *World Stem Cell Report*. Wellington, FL: Genetic Policy Institute.

> The Genetics Policy Institute sponsors a yearly conference, the World Stem Cell Summit, which it claims to be "the largest interdisciplinary, networking meeting of stem cell science and regenerative medicine stakeholders." The meeting focuses on topics such as discovery, translation

and clinical trials, innovation, regenerative services, financial issues, and hot topics and future trends. A summary of the conference speeches, reports, and other activities is then published as *World Stem Cell Report* for each year.

Javitt, Gail H., Kristen Suthers, and Kathy Hudson. 2005. "Cloning: A Policy Analysis." Genetics and Public Policy Center. http:// www.dnapolicy.org/images/reportpdfs/Cloning_A_Policy_Analysis_Revised.pdf. Accessed on January 19, 2015.

This report presents, primarily, the results of a 2004 survey of 4,834 Americans about their knowledge of and attitudes about reproductive genetic technologies, including cloning. The survey found that most Americans have incomplete or incorrect information about the topic. The report is of interest and value because it also provides background information about the technology of cloning, arguments for and against cloning, federal and state laws about cloning, and international cloning policy.

Johnson, Judith A. 2002. "Human Cloning." Congressional Research Service. http://fpc.state.gov/documents/organization/ 9666.pdf. Accessed on January 19, 2015.

This report was issued shortly after some of the earliest research on human cloning was being announced and soon after the election of President George W. Bush and his consequent announcement on U.S. policy on the federal funding of stem cell research. The short report consists of five sections, a general introduction, ethical and social issues related to the cloning of humans, a history of federal policy with regard to human embryo research, Bush administration policy about human embryo research, and legislative efforts dealing with embryo research.

Johnson Judith A., and Erin Williams. 2006. "Human Cloning." Congressional Research Service. http://crs.wikileaks-press.org/ RL31358.pdf. Accessed on January 19, 2015.

This report by the Congressional Research Service was originally not released to the general public, but was obtained by the hacking program known as WikiLeaks, which then released the report to the general public. The report was written partly in response to the report that South Korean researcher Hwang Woo-suk had falsified extensive portions of his research on human cloning and had been censured by the South Korean government and the Seoul National University, where he was employed. The report discusses the circumstances of Hwang's misbehavior and then updates an earlier report (see Johnson 2002) on U.S. federal policy on stem cell research and related topics.

Kass, Leon, ed. 2002. *Human Cloning and Human Dignity: The Report of the President's Council on Bioethics.* President's Council on Bioethics. New York: Public Affairs.

This report was prepared by the newly created President's Council on Bioethics, which was charged with studying and reporting on the ethical issues associated with human cloning. The report is one of the standard works in the field of cloning ethics and relies on extensive input from professionals in a wide variety of fields associated with cloning and ethics.

Kuppuswamy, Chamundeeswari, et al. 2007. "Is Human Reproductive Cloning Inevitable: Future Options for UN Governance?" Biodiplomacy Programme of the United Nations University. http://archive.ias.unu.edu/resource_centre/Cloning_9.20B.pdf. Accessed on January 19, 2015.

This report arose out of conflict among U.N. members in the attempt to adopt a joint policy on cloning, with the primary issue centering on the relevant risks and benefits of reproductive and therapeutic cloning. The report contains sections on the technology of cloning, cloning ethics, international governance of cloning, and future options for the international governance of cloning.

National Research Council, et al. 2010. *Final Report of the National Academies' Human Embryonic Stem Cell Research Advisory Committee and 2010 Amendments to the National Academies' Guidelines for Human Embryonic Stem Cell Research.* Washington, DC: National Academies Press.

> This report was inspired by the election of President Barack Obama and his issuance of Executive Order 13505, which relaxed restrictions on federal funding of stem cell research. Included with the executive order was the request that the National Institutes of Health begin a reconsideration of its policies and practices with regard to the use of stem cells for the cloning of human embryos. This report summarizes the results of that internal review.

Nordic Committee on Bioethics and Nordic Council of Ministers. 2000. "The Ethical Issues in Human Stem Cell Research: Report from a Workshop." Copenhagen: Nordic Council of Ministers.

> This report is an example of the many reports prepared by governmental bodies other than the U.S. federal government. In addition to such U.S. reports, many individual states, foreign nations, and regional associations have held meetings and issued reports on cloning and stem cell research. Most such reports review the relevant science and technology of stem cell research at the time of the report, with a review of the ethical issues involved in such research and the implications of these factors for state, national, regional, or international policy.

President's Council on Bioethics. 2004. "Monitoring Stem Cell Research." Washington, DC: President's Council on Bioethics. https://bioethicsarchive.georgetown.edu/pcbe/reports/stemcell/. Accessed on January 19, 2015.

> This report was issued as part of the President's Council on Bioethics original charge to study and make recommendations about U.S. policy on stem cell research. The report consists of four major parts, the first of which clarifies and explains existing federal policy about stem

cell research. The second part provides an overview of the ethical and policy debates about stem cell research that developed over the decade preceding the report. The third part reviews recent developments in the science and technology of stem cell research. The fourth section suggests some ways in which a large share of the public can begin to understand and become involved in the discussion about stem cell research in the United States.

Shimabukuro, Jon O. 2007. "Background and Legal Issues Related to Stem Cell Research." Congressional Research Service. Washington, DC: Congressional Research Service.

This report is a follow-up on a similar report by Duffy (listed earlier) on the status of stem cell research in the United States, with special attention to the many bills on the topic introduced in the Congress during the 110th session.

Internet

Anderson, Scott C. "A Baby's Hair." Science for People. April 16, 2004. http://www.scienceforpeople.com/Essays/baby_hair.htm. Accessed on July 26, 2014.

This article is a delightful description of some of the earliest experiments conducted on cloning by German embryologist Hans Spemann in the early 20th century.

"Animal Biotechnology." Bioscience Topics. http://www.aboutbioscience.org/topics/animalbiotechnology. Accessed on January 19, 2015.

This Web site deals with a variety of topics associated with the general topic of animal biotechnology, cloning being one of them. It discusses the history and technology of animal cloning, along with current progress in the field and social, ethical, and other issues related to animal cloning.

Byrne, James A. 2014. "Developing Neural Stem Cell-Based Treatments for Neurodegenerative Diseases." *Stem Cell Research &*

Therapy. 5:72 (online article). http://link.springer.com/ article/10.1186%2Fscrt461. Accessed on January 20, 2015.

    Byrne provides a succinct, well-written, easily understandable explanation of the way that stem cells might be used for the treatment of neurodegenerative diseases such as Alzheimer's disease.

"Cloned Animals." Royals Society for the Prevention of Cruelty to Animals (RSPCA). http://www.rspca.org.uk/advicean-dwelfare/laboratory/biotechnology/clonedanimals. Accessed on January 19, 2015.

    The RSPCA explains the process by which animals are cloned and then tells what the organization thinks about the process and its potential applications. (It believes that animals are being cloned without consideration to their health and welfare.) The Web site is particularly valuable for its links to other sources on the cloning of animals.

"Clone Food Products? We Just Call Them Apples." Fruit Growers News. http://fruitgrowersnews.com/index.php/magazine/article/ Cloned-Food-Products-We-Just-Call-Them-Apples. Accessed on January 19, 2015.

    This article argues that growers have been cloning plants for centuries, without any concerns about the safety of food produced from those plants. It says that the same attitude should be applied to food products obtained from cloned animals.

"Cloning." Encyclopedia.com. http://www.encyclopedia.com/ topic/cloning.aspx. Accessed on January 18, 2015.

    Encyclopedia.com is an online electronic publication of Cengage Learning. It consists of articles taken from a number of important print encyclopedias, such as the Columbia Encyclopedia and publications from Oxford University Press. This section comes from Chapter 8 of *Genetics and Genetic Engineering*, by Barbara Wexler (Information Plus® Reference Series, 2007).

"Cloning." European Food Safety Authority. http://www.efsa.
europa.eu/en/topics/topic/cloning.htm. Accessed on January
19, 2015.

> This Web site summarizes the regulatory decisions and
> policy on the cloning of animals in Europe and the role
> of the EFSA in administering and carrying out these rules
> and policies.

"Cloning." Learn.Genetics. Genetic Science Learning Center.
http://learn.genetics.utah.edu/content/cloning/. Accessed on July
30, 2014.

> This Web page provides a good general introduction to
> the topic of cloning with separate sections on the history
> of cloning, an interactive experiment on cloning, an inter-
> active quiz on the subject, a discussion of the reasons for
> cloning, and a review of some cloning myths.

"Cloning." National Human Genome Research Institute. April 28,
2014. http://www.genome.gov/25020028#al-2. Accessed on July
30, 2014.

> This Web page provides a very general introduction to the
> topic of cloning, with additional sections on the pros and
> cons of cloning technology.

"Cloning." The New Scientist. http://www.newscientist.com/searc
h?doSearch=true&query=cloning. Accessed on January 19, 2015.

> This Web page lists more than 900 articles on the topic of
> cloning that have appeared in *The New Scientist* over the
> past decade and a half.

"Cloning." New York Times. http://topics.nytimes.com/top/news/
science/topics/cloning/index.html. Accessed on January 18, 2015.

> The New York Times maintains and provides a listing of
> articles on a variety of important current topics, clon-
> ing being one of them. This Web site lists all of the ar-
> ticles about cloning that have appeared in the newspaper
> since 1984.

"Cloning." Scientific American. http://www.scientificamerican. com/topic/cloning/. Accessed on January 18, 2015.

This Web site contains a number of articles that have appeared in the *Scientific American* magazine on various aspects of cloning, such as the possibilities of cloning endangered and extinct species, technical advances in the production of stem cells, legal and policy issues related to cloning research, and interesting experiments that amateurs can conduct on cloning.

"Cloning." The Scientist. http://www.the-scientist.com/?articles. list/tagNo/435/tags/cloning/pageNo/3/. Accessed on January 18, 2015.

The Scientist magazine provides this Web site, listing articles that have appeared on a variety of scientific topics, cloning being one of them. Some sample topics are New Ways to Make Embryonic Cells, Chinese Miracle Pig Cloned, Clone Wars, A New Breed (produced by cloning), Chinese Cloning Researcher Arrested, and Korean Stem Cell Film Tops Box Office.

"Cloning Magazine." http://cloningmagazine.com/index.html. Accessed on January 18, 2015.

This Web site is designed to report on "current scientific breakthroughs in the expanding genetics and cloning business" during 2015. At the beginning of the year, it provided background articles on topics such as the technology and issues related to research with stem cells, cloning research, and the generation of body parts by cloning technology.

"Cloning News." Science Daily. http://www.sciencedaily.com/ news/plants_animals/cloning/. Accessed on January 18, 2015.

Science Daily is a Web site that provides prompt, accurate, and comprehensive information on a variety of scientific topics written largely for the general public. The Web site includes a section on cloning that contains the most important and most recent articles on cloning research and related topics.

"Cloning's Historical Timeline." National Science Teachers Association. http://www.nsta.org/images/news/legacy/scope/0603/cloningtimeline.pdf. Accessed on January 19, 2015.

This Web site provides one of the most complete timelines of the history of cloning available on the Internet. None of the Web sites on which it is based is available any longer. This timeline has the disadvantage of ending in 2005, so it should be supplemented with a more recent timeline, such as "Send in the Clones," listed later.

Connors, Russell B., Jr. "The Ethics of Cloning." St. Anthony Messenger. http://www.americancatholic.org/Messenger/Mar1998/Feature2.asp. Accessed on January 19, 2015.

This essay provides an opinion on the ethical implications of cloning from the perspective of a Roman Catholic writer and teacher. He concludes that "[t]he cloning of plants and animals for the purposes of greater productivity may well fit with our charge to be stewards of the earth and instruments of God's creation. But it would ultimately benefit none of the inhabitants of the earth if cloning were to take place at the expense of plant and animal diversity."

Cottrell, Sariah, Jamie L Jensen, and Steven L Peck. 2014. "Cheetahs, Mammoths, and Neanderthals." *Life Sciences, Society and Policy.* http://www.lsspjournal.com/content/10/1/3. Accessed on January 19, 2015.

This article considers the ethical implications of cloning three categories of animals, represented by cheetahs (endangered species), mammoths (extinct species), and Neanderthals (humanoid species). The authors conclude that cloning endangered species is probably ethically acceptable, cloning of extinct species may or may not be ethically acceptable, and the cloning of humonid species is almost certainly not ethically acceptable.

Freudenrich, Craig. "How Cloning Works." How Stuff Works. http://science.howstuffworks.com/life/genetic/cloning.htm. Accessed on January 19, 2015.

This article very nicely introduces and explains the process of cloning. Various sections deal with the cloning of plants, animals, and humans, along with links to other "How Stuff Works" articles on related topics.

"General Q&A." U.S. Food and Drug Administration. http://www.fda.gov/AnimalVeterinary/DevelopmentApprovalProcess/GeneticEngineering/GeneticallyEngineeredAnimals/ucm113605.htm. Accessed on January 19, 2015.

This Web site focuses first on an explanation of the process by which animals are cloned. It concentrates primarily on regulatory issues related to the cloning of animals in the United States, such as which agencies are responsible for the regulation of research and development, what rules may control cloning procedures, and what agencies to contact for additional information on the topic.

"Genetically Engineered Salmon." U.S. Food and Drug Administration. http://www.fda.gov/AnimalVeterinary/Development ApprovalProcess/GeneticEngineering/GeneticallyEngineered-Animals/ucm280853.htm. Accessed on January 19, 2015.

This Web site reviews the development of the AquAdvantage Salmon, whose producers have applied for permission to sell the cloned fish domestically. It describes the regulatory process by which approval is to be gained and the current status of that process.

Hinrichs, Katrin. "Cloning of Domestic Animals." The Merck Veterinary Manual. http://www.merckmanuals.com/vet/management_and_nutrition/cloning_of_domestic_animals/overview_of_cloning_of_domestic_animals.html. Accessed on January 19, 2015.

This highly respected reference manual contains sections on the cloning of domestic animals dealing with a general overview of the topic, technical aspects of cloning, status of the cloning of domestic animals, rationale for cloning, and controversies about cloning.

"Human Cloning Laws: 50 State Survey." Bioethics Defense Fund. May 19, 2011. http://bdfund.org/wordpress/wp-content/uploads/2012/07/CLONINGChart-BDF2011.docx.pdf. Accessed on July 31, 2014.

This report provides one of the most complete summaries of the status of cloning legislation in the United States as of 2011. It lists eight states that prohibit cloning for any purpose whatsoever, four states that prohibit funding for cloning projects of any kind, ten states that prohibit human cloning but allow therapeutic cloning, five states that allow funding for such research, and two states that specifically allow doctors to refuse to participate in any type of cloning research.

Jabr, Ferris. 2013. "Will Cloning Ever Save Endangered Animals?" Scientific American. http://www.scientificamerican.com/article/cloning-endangered-animals/. Accessed on January 18, 2015.

The author describes the process by which endangered animals can be cloned and discusses some of the advantages and problems associated with using this technology in the real world. The program being developed to save some of Brazil's endangered species is described in some detail.

King, Nancy M. P., and Jacob Perrin. 2014. "Ethical Issues in Stem Cell Research and Therapy." *Stem Cell Research and Therapy*. 5(4) (online article). http://stemcellres.com/content/5/4/85. Accessed on January 20, 2015.

This review article provides an excellent overview of the status of stem cell research and therapy as of 2014, along with a summary of the ethical issues related to all aspects of that research and its applications.

Klinkenborg, Verlyn. "Closing the Barn Door after the Cows Have Gotten Out." New York Times. http://www.nytimes.

com/2008/01/23/opinion/23wed4.html?_r=2&scp=1&
sq=Verlyn%20Klinkenborg%20closing%20the%20barn%20
door&st=cse&. Accessed on January 19, 2015.

The writer explains that she is opposed to eating food
from cloned animals not for safety or health reasons, but
because of the harm it poses to agricultural systems.

Li, Jun, et al. 2014. "Advances in Understanding the Cell Types
and Approaches Used for Generating Induced Pluripotent
Stem Cells." *Journal of Hematology & Oncology.* 7:50 (online article). http://www.jhoonline.org/content/7/1/50. Accessed on
January 20, 2015.

The authors provide a comprehensive and very readable
account of the development of induced pluripotent stem
cells and their present role in research and therapy.

Moon, Seongwuk, and Seong Beam Cho. 2014. "Differential
Impact of Science Policy on Subfields of Human Embryonic
Stem Cell Research." *PloS ONE.* 9(4): e86395.

Governmental support has varied widely in the two
decades during which it has been a significant area of
research. In the United States, for example, this line of
research was largely discouraged for at least eight years
during the administration of President George W. Bush.
During the same period, stem cell research was encouraged and supported by federal funding in other nations
of the world. The authors of this survey studied the effect of varying governmental policies about the funding of stem cell research between 1998 and 2008 and
divided their study into three major areas: derivation,
differentiation, and applications. They found some very
interesting and very significant differences in the way
that national policies affected development of research
in each of these three areas from country to country.

Pappas, Stephanie. "Human Cloning? Stem Cell Advance Re-
ignites Ethics Debate." Live Science. http://www.livescience.

com/34487-human-cloning-stem-cell-ethics.html. Accessed on January 19, 2015.

A recent discovery in methods for producing stem cells for the cloning of animals has renewed the debate over human reproductive cloning, according to this writer, and she raises once again some of the social and ethical issues posed by the potential to produce human clones.

Polak, Julia. "Stem Cells & Therapeutic Cloning." Medical Research Council [United Kingdom]. http://resources.schoolscience.co.uk/MRC/cloning/page1.html. Accessed on January 19, 2015.

This chapter is part of the Medical Research Council's (MRC) Research Updates project, designed to bring information about current developments in medicine and biology to the general public. This essay provides an excellent overview of the topic of therapeutic cloning and stem cell research with sections on the technology involved, the potential benefits of stem cell research, issues related to stem cell studies, and a review of current UK law on cloning and stem cell research.

"Regenerative Medicine." 2006. Stem Cell Information. National Institutes of Health. http://stemcells.nih.gov/info/scireport/pages/2006report.aspx. Accessed on January 21, 2015.

This booklet consists of a number of chapters written by experts in the field of stem cell research and regenerative medicine. The information is clearly written and accessible to young adults, supplemented by a very large number of important and valuable references. The chapters deal with a general introduction to stem cell research; bone marrow stem cells; the nervous system; experimental gene therapies; intellectual property of human pluripotent stem cells; cardiac repair; diabetes treatment; alternate methods for preparing pluripotent stem cells; cancer; induced pluripotent stem cells; and bone structure, function, and formation. The booklet was updated in 2010 and 2011 to include new information about recent stem cell research.

Sass, Reuben G. 2014. "Bringing Prosocial Values to Translational, Disease-Specific Stem Cell Research." *BMC Medical Ethics*. 15(16) (online article). http://www.biomedcentral.com/1472-6939/15/16. Accessed on January 20, 2015.

> The author proposes a system whereby prizes or awards are offered that will "spur innovation in delivery of care, promoting 'prosocial' values of transparency, equity, patient empowerment, and patient-provider and inter-institutional collaboration." He describes such a program within the context of using disease-specific stem cell therapies for severe, high-maintenance, chronic disorders.

Scott, Christopher Thomas, and Irving L. Weissman. 2008. "Cloning." The Hastings Center. http://www.thehastingscenter.org/Publications/BriefingBook/Detail.aspx?id=2158. Accessed on January 17, 2015.

> This Web page is taken from *Birth to Death and Bench to Clinic: The Hastings Center Bioethics Briefing Book for Journalists, Policymakers, and Campaigns*, edited by Mary Crowley (Garrison, NY: The Hastings Center, 2008). It provides a very good general introduction to the subject of cloning, including a historical review of the topic, bioethical considerations, legal and policy issues, glossary, and links to other Web sites on the topic.

"Send in the Clones." BBC Future. http://ichef.bbci.co.uk/ww-features/original/images/live/p0/0p/kb/p00pkbsv.jpg. Accessed on January 19, 2015.

> This Web page provides a very interesting timeline on the history of cloning since 1960. It is among the most up-to-date in its treatment of cloning history, but has little about the early history of cloning. It should, therefore, be viewed in association with a Web site that does so, such as "Cloning's Historical Timeline," listed earlier.

Shreeve, Jamie. "Species Revival: Should We Bring back Extinct Animals?" National Geographic. http://news.nationalgeographic.

com/news/2013/03/130305-science-animals-extinct-species-re-vival-deextinction-debate-tedx/. Accessed on January 19, 2015.

> This article provides an excellent and complete review of the current status of research on the cloning of extinct animals with a review of the pros and cons of attempting to recover some specific species that have gone extinct by cloning.

Sohn, Emily. "Animal Clones: Double Trouble?" Student Science. https://student.societyforscience.org/article/animal-clones-double-trouble. Accessed on January 18, 2015.

> This Web site provides articles on a variety of science-related articles written for young adults with additional resources related to the topic. This article deals with the technology of animal cloning, with a review of some of the benefits of animal cloning, as well as some of the problems associated with the procedure.

Sterns, Robin K. "Double or Nothing." Santa Clara University Markkula Center for Applied Ethics. http://www.scu.edu/ethics/publications/submitted/sterns/doublenothing.html. Accessed on January 19, 2015.

> For an article in the *Santa Clara Magazine*, the author interviewed experts on the Santa Clara campus in the fields of biology, medicine, religious studies, patent law, engineering, technology, agriculture, and ethics to elicit their views on the cloning of animals. This Web page summarizes the responses she received in these interviews.

Wilmut, Ian. "Produce Mammoth Stem Cells, Says Creator of Dolly the Sheep." The Conversation. http://theconversation.com/produce-mammoth-stem-cells-says-creator-of-dolly-the-sheep-16335. Accessed on January 18, 2015.

> Wilmut, who directed the cloning of Dolly, the sheep, believes that it should now be possible to clone the extinct woolly mammoth. In this article, he explains how that process could be carried out (it would be different from the one by which Dolly was produced) and what the possible reasons for doing, or not doing, this research might be.

Introduction

The topic of cloning often sounds like one of the "hot top-ics" in today's world. But experiments related to cloning have been conducted for more than a century. Highly sophisticated modern technology for producing identical copies of mol-ecules, cells, or organisms owes a great deal to this long his-tory of cloning research. Some of the most important events in this history are detailed in this chapter. (For additional ref-erences on specific aspects of the chronology of cloning, see Tissue Culture and Its History [http://agridr.in/tnauEAgri/eagri50/GPBR311/lec02.pdf, accessed on January 1, 2015]; History of Plant Tissue Culture [http://www.sciencebeing.com/2014/03/history-of-plant-tissue-culture/, accessed on January 1, 2015]; Wright, Mary V. "Cloning: A Select Chro-nology." Congressional Research Service [http://www.cnie.org/nle/crsreports/03Apr/RL31211.pdf, accessed on January 1, 2015]; Fairbanks, Stephen D., ed. *Cloning: Chronology, Ab-stracts, and Guide to Books.* New York: Nova Science Publishers, 2004; Solter, Davor. 2006. "From Teratocarcinomas to Em-bryonic Stem Cells and beyond: A History of Embryonic Stem Cell Research." *Nature Reviews Genetics.* 7(4): 319–327.)

James Symington poses with puppies cloned from a German Shepherd that reportedly took part in the search-and-rescue effort after the 9/11 terrorist attacks. Symington won an essay contest in 2008 to clone his dog Trakr for free. (AP Photo/Damian Dovarganes)

ca. 1770 The "Williams" (later "Bartlett") pear is first cloned in England.

1831 Scottish botanist Robert Brown first identifies and names the nucleus of a cell and suggests that it has an essential role in the process of fertilization.

1838 German botanist Matthias Jakob Schleiden proposes the cell theory of plants.

1839 German physiologist Theodor Schwann extends the cell theory to include animals.

1855 Polish-German embryologist Robert Remak proposes the theory that embryos grow and develop as the result of the division of preexisting cells.

1868 German biologist Ernst Haeckel uses the term *Stammzelle* ("stem cell") to refer to primitive cells from which all multicellular organisms develop.

1869 Swiss physician and biologist Johannes Friedrich Miescher discovers a previously unknown phosphorus-rich compound that he calls nuclein. The compound is later found to be a nucleic acid, although its biological function is not understood for many years.

1876 German zoologist Oscar Hertwig suggests that fertilization occurs when the sperm penetrates an egg.

ca. 1880 The "Delicious" apple is first cloned in an unknown Iowa orchard.

1880s German biologists August Weismann and Wilhelm Roux independently and simultaneously develop the germ plasma theory of plant growth and development.

1885 German embryologist Hans Adolf Eduard Driesch performs what many consider to be the first cloning experiment in history when he separates the two cells in a two-cell sea urchin embryo simply by shaking it. The two individual cells then grow normally into adult sea urchins, which are, however, somewhat smaller than normal.

1895 French biologist Yves Delage outlines some of the basic principles involved in nuclear transfer experiments, later realized more fully by Hans Spemann and German physiologist Jacques Loeb, a contemporary of Delage.

1902   Austrian botanist Gottlieb Haberlandt conducts early experiments on tissue culture and proposes the idea of totipotentiality, suggesting that all cells in a plant contain the potential of developing into a complete plant.

1902   German embryologist Hans Spemann clones a salamander by dividing the two cells in a two-cell embryo and allowing each cell to grow independently. He conducts the experiment by tying a strand of his baby daughter's hair around the middle of the two-cell embryo.

1902   University of Kansas bacteriologist Marshall Barber invents the micropipette, a device made of a capillary tube with a very small opening at one hand. Barber used the device to capture individual bacteria for study, although they are also widely used today to extract the nucleus from a cell for transfer to a second enucleated cell.

1903   Herbert J. Webber, a plant physiologist at the U.S. Department of Agriculture, suggests the term *clone* to describe a method by which plants and plant parts are propagated asexually. The word comes from the Greek word *klon*, meaning a twig or other plant part that is removed for the purpose of propagation.

1904   German botanist Emil Hannig conducts the first plant tissue culture experiments using embryonic cells from plants.

1908   German botanist Siegfried Simon attempts to grow complete plants from the callus of parent plants, a project in which he is only partially successful.

1922   German botanist Walter Kotte and American botanist William J. Robbins independently and almost simultaneously explore the possibility of cloning plants from root tips.

1926   Dutch botanist Fritz Went discovers the first plant auxin, indoleacetic acid (IAA).

1930s   Danish-German embryologist Joachim Hämmerling carries out some of the earliest nuclear transfer experiments in which he inserts the nucleus of one variety of the giant unicellular alga *Acetabularia* into a second enucleated individual of the same genus. The second individual develops into a mature plant with the characteristics of the plant from which the nucleus was transferred.

1932   American botanist Philip White begins an extended series of experiments to determine the effect of culture medium on the cloning of plants.

1934   French botanist Roger Jean Gautheret produces some of the "first true plant tissue cultures" by selecting a judicious combination of culture media.

1935   Russian biochemist Andrei Nikolaevitch Belozersky isolates DNA in the pure state for the first time.

1938   Spemann publishes his book *Embryonic Development and Induction* in which he reviews and summarizes the results of his experiments of the preceding two decades and outlines a protocol for future experiments on somatic cell nucleus transfer. The book also contains a description of a "fantastical experiment" that was eventually realized as somatic cell nuclear transplantation (SCNT).

1944   Oswald Avery demonstrates that DNA is the carrier of hereditary information.

1952   American researchers Robert Briggs and Thomas J. King successfully transfer blastula cells into the enucleated cells of the northern leopard frog (*Lithobates pipiens*, formerly known as *Rana pipiens*). The experiment results in the production of tadpoles that are clones of the original organisms with which the nuclear transfer was conducted.

1953   British chemist Francis Crick and American biologist James Watson discover the three-dimensional structure of DNA.

1958   English embryologist Sir John Gurdon produces living tadpoles of the African clawed frog (*Xenopus laevis*), using nuclear transplantation technology. The tadpoles are allowed to grow to maturity, after which they are able to reproduce as successfully as their nonengineered cousins.

1961   Canadian researchers James Till and Ernest Mc-Culloch at the Ontario Cancer Institute demonstrate the existence of stem.

1963   Chinese embryologist Tong Dizhou produces the first cloned fish, a type of Asian carp, *Rhodeus sinensis*.

1967   The first ligase is discovered almost simultaneously at five different laboratories.

1969   American researchers Jonathan Beckwith, Lawrence Eron, and James Shapiro make the first discovery of a gene.

1970   American microbiologists Hamilton O. Smith, Thomas Kelly, and Kent Wilcox discover the first restriction enzyme (restriction endonuclease).

1973   Tong produces the first interspecics clone by transplanting nuclei from two related, but distinct, species, the European crucian carp (*Carassius carassius*) and an Asian carp (*Rhodeus sinensis*).

1974   The U.S. Congress passes the National Research Act, one section of which establishes the National Commission for the Protection of Human Subjects in Biomedical and Behavioral Research, whose purpose is to study issues related to fetal research. The act also places a moratorium on fetal research until the commission has reported back to Congress. For most practical purposes, the act brings to a halt the federal funding of most research involving human fetuses for more than three decades.

1978   The Genentech Corporation announces the first commercial production of insulin, which it calls Humulin, using genetic engineering technology.

1981   The University of California, San Francisco, researcher Gail Martin isolates stem cells from a mouse embryo and demonstrates that they are pluripotent. The University of Cambridge researchers, Martin Evans and Matthew Kaufman, report similar results almost simultaneously with those of Martin.

1982 The U.S. Food and Drug Administration (FDA) approves Humulin for use with humans for the treatment of diabetes mellitus.

1985 Congress passes the Health Research Extension Act, which prohibits any federally funded research that causes harm to a fetus, essentially reinforcing and extending the existing federal prohibition on fetal research.

1986 Swiss biologist Karl Illmensee and American biologist Peter C. Hoppe claim to have cloned three mice. The claim is later found to have been false and, perhaps, falsified.

1986 Danish embryologist Steen Willadsen successfully performs the first cloning of a mammal, a sheep.

1993 President Bill Clinton and the Congress appear to reverse course on U.S. policy on the funding of fetal research. Clinton directs the Secretary and Health and Human Services to lift the ban on the federal funding of therapeutic cloning fetal research, and Congress passes the National Institutes of Health Revitalization Act, which confirms Clinton's new policy. Clinton also appoints a Human Embryo Research Panel to make recommendations about federal funding for therapeutic cloning research. (Also see 1994.)

1994 The Human Embryo Research Panel, appointed by President Clinton in 1993, recommends that federal funding for therapeutic research be funded by the U.S. government. In a change of heart, however, Clinton rejects the panel's recommendation.

1995 With a Republican majority in both houses for the first time since the 1950s, the U.S. Congress votes to reverse President Clinton's policies on cloning research by passing the Dickey Amendment, which prohibits the federal funding for fetal cloning research of any kind whatsoever. The amendment remains in effect until 2011.

1996   Ian Wilmut and his colleagues at the Roslin Institute in Scotland successfully clone a sheep that they name Dolly using SCNT for the first time in such research.

1997   California becomes the first state to enact a cloning law when the legislature passes a bill banning human reproductive cloning and creating a commission to determine what state policy should be about cloning once that law expires.

1997   The Council of Europe adopts the Convention for the Protection of Human Rights and Dignity of the Human Being with regard to the Application of Biology and Medicine, generally regarded as the first international statement protecting the rights of the fetus and prohibiting certain forms of cloning.

1998   Research teams led by James A. Thomson at the University of Wisconsin, Madison, and John D. Gearhart at the Johns Hopkins University isolate human embryonic stem cells. Both teams successfully maintain the cells for an extended period of time in vitro and then allow them to develop through about 20 passages.

1998   The Additional Protocol on the Prohibition of Cloning Human Beings to the Convention for the Protection of Human Rights and Dignity of the Human Being with regard to the Application of Biology and Medicine is adopted by the Council of Europe. The protocol provides more specific protection to the human fetus in scientific research stating that "[a]ny intervention seeking to create a human being genetically identical to another human being, whether living or dead, is prohibited."

1999   The first male mammal, a mouse named Fibro, is successfully cloned.

1999   A rhesus macaque, called Tetra, is cloned.

2000   American billionaire John Sperling and his wife Joan Hawthorne form a corporation called Genetic Savings & Clone for the purpose of cloning pets for individuals who can afford to pay the cost of the procedure. Their first successful procedure results in the birth of a clone cat called Copy Cat (or C.C., or CC) a year later.

2001 The first domestic cat, called CC, is cloned.

2001 The first endangered animal, a gaur, is cloned.

2001 In a speech to the nation, President George W. Bush announces new regulations regarding the permitted status of federal funding for stem cell research. The order bans all such research except for 64 lines of stem cells already in existence. It later develops that only about a third of those lines are of any value to researchers.

2002 The cloning company Clonaid announces the first cloning of a human child. No reliable evidence for the existence of such a child is ever adduced.

2002 Austrian-American geneticist Konrad Hochedlinger and Polish-American geneticist Rudolf Jaenisch successfully produce fertile mice with nuclear transfer technology using fully differentiated lymphocyte cells.

2002 The U.S. Patent and Trademarks Office issues a patent to the University of Missouri for a procedure for the cloning of mammals. Critics point out that the procedure could also be used for the cloning of humans at some time in the future although the university disavows any intention ever to do so.

2003 The first extinct animal, a Pyrenean ibex (or bucardo), is cloned.

2004 South Korean theriogenologist Hwang Woo-suk announces the creation of human embryonic stem cells by cloning technology. At the time, Hwang is already world famous for his accomplishments in the field of cloning, such as the reported cloning of cows as early as 1999. (But see 2006 and 2009.)

2004 The New Jersey state legislature becomes the first state to authorize a state-funded stem cell research, creating the Stem Cell Institute of New Jersey. Voters decline to provide funding for the institute, however, and legislative plans for the facility are never realized (at least as of 2015).

2005   Representative Michael Castle (R-DE) introduces the Stem Cell Research Enhancement Act into the 109th Congress. The bill passes both houses of Congress but is vetoed by President George W. Bush. (Also see 2007 and 2009.)

2006   Genetic Savings & Clone goes out of business. Shortly thereafter, a group of South Korean researchers establishes the Sooam Biotech Research Foundation for the purpose of developing the technology for cloning dogs, cats, and other pets, as well as the cloning of other animals for research purposes.

2006   A report produced by investigators at Seoul National University concludes that reports of cloning successes by Hwang Woo-suk were entirely fabricated. (See 2009.)

2006   Researchers at Kyoto University and the University of California, San Francisco, successfully produce induced pluripotent stem cells (iPSCs) for the first time. Leader of the project, Shinya Yamanaka, is awarded the 2012 Nobel Prize in Physiology or Medicine for the achievement.

2007   The Stem Cell Research Enhancement Act is reintroduced, passed by both House and Senate, and again vetoed by President George W. Bush. Bush's veto is upheld in a later Congress vote. (Also see 2005 and 2009.)

2009   A South Korean court convicts Hwang Woo-suk of falsifying research reports and embezzling government research funds. He is sentenced to a two-year prison term, which is then suspended. Hwang later continues his cloning research and eventually produces significant results in the cloning of pets and research animals.

2009   The Stem Cell Research Enhancement Act is reintroduced into the 111th Congress but is never acted upon.

2009   President Barack Obama issues Executive Order 13505, which reverses federal policy on the federal funding of human embryonic research that dates back more than two decades.

2011   The U.S. Court of Appeals for the District of Columbia overturns the Dickey Amendment, first enacted in 1995.

2011   Representative Diana DeGette (D-CO) introduces the Stem Cell Research Advancement Act to extend federal funding of stem cell research. The act is never acted on by either the House or Senate. (Also see 2013.)

2012   The Brasilia Zoological Garden and EMBRAPA, Brazil's national agricultural research agency, announce a new program to clone eight of the country's most endangered animals, the maned wolf (*Chrysocyon brachyurus*), jaguar (*Panthera onca*), black lion tamarin (*Leontopithecus chrysopygus*), bush dog (*Speothos venaticus*), coati (*Nasua*), collared anteater (*Tamandua tetradactyla*), gray brocket deer (*Mazama gouazoubira*), and bison (*Bison*).

2012   John Gurdon and Shinya Yamanaka are awarded the 2012 Nobel Prize for Physiology or Medicine for their research in the field of regenerative medicine. (See 1958 and 2006.)

2013   Researchers at Oregon Health Sciences University announce that they have successfully cloned human embryonic stem cells produced by SCNT.

2013   Representative Diana DeGette (D-CO) reintroduces the Stem Cell Research Advancement Act, which again fails to receive consideration in either the House or Senate.

2013   Researchers at the Lazarus Project report that they have successfully cloned embryos from the extinct gastric-brooding frog, *Rheobatrachus silus*. Although the embryos survive only a short time, they represent an important first step in the de-extinction of a species that died out in the 1980s.

2014   Going one step beyond results announced in 2013 (See 2013), two teams of researchers report that they have produced human embryonic stem cell lines from embryos cloned from adult cells. The advance means that it should be possible to develop therapies designed for specific individuals using their own unique stem cells.

2014   Researchers in China announce that they have begun producing cloned pigs "on an industrial level," with 500 such pigs having been born in the past year and even larger number scheduled for future years. The success rate of the cloning procedure ranges between 70 and 80 percent. The pigs are being used not for food but for drug research.

2015   Scientists at Harvard University move one step forward in the process of cloning the extinct woolly mammoth by inserting some sequences of the animal's DNA into an elephant genome.

Glossary

Introduction

As is the case with most technical issues, discussions of cloning involve the use of a great many specialized terms with which one must become familiar to understand the topic at hand. The glossary provided here includes many, but by no means all, of those terms. Additional sources of excellent glossaries dealing with cloning terms can be found at http://www.nap.edu/openbook.php?record_id=10285&page=259; http://www.leaderu.com/science/stemcellcloning_glossary.html; and http://www.fda.gov/animalveterinary/safetyhealth/animalcloning/ucm124953.htm. A particularly appealing introduction to the vocabulary of cloning is "Talking Glossary of Genetic Terms," provided by the National Human Genome Research Institute at http://www.genome.gov/Glossary/, which provides both audio and verbal explanations of cloning terminology provided by researchers at the Institute.

Adult stem cell A cell taken from a mature individual that has the potential to become pluripotent or multipotent and differentiate into other types of cell.

Asexual reproduction A form of reproduction that takes place without the fusion of gametes.

Bacteriophage A virus that infects bacteria. Also known simply as a phage.

Blastocoele   The inner fluid-filled cavity in a blastula. *See* **Blastula**

Blastocyst   A thin-walled hollow structure that appears early in embryonic development. A blastocyst contains a cluster of cells called the inner cell mass from which the embryo arises. An outer layer of cells gives rise to the placenta and other tissues needed for fetal development.

Blastomere   One of the cells formed by the cleavage of a fertilized egg.

Blastula   A hollow sphere of cells surrounding an inner fluid-filled cavity formed during an early stage of embryonic development in animals. The cells are known as blastomeres, and the interior cavity is called the blastocoele.

Chimera   A piece of DNA or an animal produced by combining the DNA of two different organisms.

Clone   A molecule, cell, or organism that is an exact copy of some other molecule, cell, or organism.

De-extinction   The process of attempting to produce animals, often by the process of cloning, that have become extinct in the natural world.

Deoxyribonucleic acid.   *See* **DNA**.

Differentiation   The process by which a simple cell, such as a zygote, matures to become a cell with a specialized function, such as a muscle or nerve cell.

DNA   A molecule consisting of a pair of strands consisting of a sugar (deoxyribose), a phosphate group, and four nitrogen bases. DNA stands for deoxyribonucleic acid.

DNA ligase.   *See* **Ligase**.

DNA marker.   *See* **Genetic marker**.

Electroporation   A process used in gene transfer in which cells are subjected to a pulse of electricity, causing pores in the cell membranes to open briefly, allowing vectors to enter the cell.

Embryonic germ cell   A cell that develops into mature gametes (sperm or eggs).

Embryonic stem cell   A cell from the inner group of cells of a blastocyst, with the potential to become all types of cells found in body tissues.

Enucleated   Without a nucleus.

Fetus   The term used to describe a developing mammal at least eight weeks after conception.

Gamete   A male or female reproductive cell that contains half the genetic material of an organism.

Gene gun   A modified air pistol used in gene transfer experiments to fire microscopic gold particles coated with DNA into a cell.

Genetic marker   A distinctive segment of DNA that can be used to identify the cells in a gene transfer experiment that have taken up foreign DNA. Certain antibiotic resistant genes are often used as markers. Also known as a marker gene, DNA marker, or reporter gene.

Host cell   In genetic engineering, a cell into which new genetic material is inserted.

Ligase   An enzyme that repairs "broken" molecules by joining two strands of the molecule. DNA ligase "repairs" broken DNA molecules by joining nucleotides in adjacent strands of a DNA molecule.

Marker gene.   *See* **Genetic marker**.

Multipotent   Capable of developing into many, but a limited number of, different types of cells.

Nitrogen base   An organic molecule that contains nitrogen and has the chemical properties of a base. Nucleic acids consist of any combination of five nitrogen bases, adenine, cytosine, guanine, and thymine (in DNA) or uracil (in RNA).

Nucleoside   A chemical structure consisting of a five-carbon sugar and a nitrogen base.

Nucleotide   A chemical structure consisting of a five-carbon sugar, a nitrogen base, and at least one phosphate group. Nucleotides are the basic units from which nucleic acids are built.

Parthenogenesis   A form of asexual reproduction in which an unfertilized egg develops without the involvement of a male gamete. Parthenogenesis may also occur with a male gamete, but no egg, but only in very rare circumstances.

Phage   *See* **Bacteriophage**.

Plasmid   A circular strand of DNA that normally occurs in bacteria but that is often used as a vector in inserting a gene into a host cell.

Plasticity   The ability of a stem cell to change into other cell types.

Pluripotent   Capable of developing into many, but not all, types of cells.

Reporter gene.   *See* **Genetic marker**.

Reproductive cloning   A procedure in which a cloned embryo is implanted into a womb with the goal of producing a live birth of a cloned organism.

Restriction enzyme   An enzyme that cuts a DNA molecule at a specific recognition site consisting of a characteristic series of nitrogen bases.

SCNT   *See* **Somatic cell nuclear transplantation**.

Somatic cell   Any cell that constitutes a part of an organism's body. A somatic cell is distinguished from a germ cell, a cell involved in the process of reproduction.

Somatic cell nuclear transfer   *See* **Somatic cell nuclear transplantation**.

Somatic cell nuclear transplantation (SCNT)   A process by which genetic material from one organism is transferred into an enucleated egg of a second organism. Also known as somatic cell nuclear transfer.

Stem cell   A type of cell capable of reproducing itself and developing into mature cells of various types.

Therapeutic cloning   The production of embryonic stem cells for treating a disease by replacing or repairing damaged tissues or organs.

Theriogenology   The branch of veterinary medicine concerned with the reproduction of animals as well as the physiology and pathology of their male and female reproductive systems.

Totipotent   Capable of developing into any type of cell.

Unipotent   Capable of developing into only one type of cell.

Vector   Any substance by which genetic material can be inserted into a host cell.

Zygote   A cell formed from the union of two gametes, a sperm and an egg.

About the Author

DAVID E. NEWTON holds an associate's degree in science from Grand Rapids Junior College, Michigan; a BA in chemistry (with high distinction); an MA in education from the University of Michigan; and an EdD in science education from Harvard University. He is the author of more than 400 textbooks, encyclopedias, resource books, research manuals, laboratory manuals, trade books, and other educational materials. He taught mathematics, chemistry, and physical science in Grand Rapids, Michigan, for 13 years; was professor of chemistry and physics at Salem State College in Massachusetts for 15 years; and was adjunct professor in the College of Professional Studies at the University of San Francisco for 10 years.

The author's previous books for ABC-CLIO include *Global Warming* (1993), *Gay and Lesbian Rights—A Resource Handbook* (1994, 2009), *The Ozone Dilemma* (1995), *Violence and the Mass Media* (1996), *Environmental Justice* (1996, 2009), *Encyclopedia of Cryptology* (1997), *Social Issues in Science and Technology: An Encyclopedia* (1999), *DNA Technology* (2009), *Sexual Health* (2010), *Science and Political Controversy* (2013), *LGBT Youth Issues Today* (2013), *GMO Food* (2014), *Wind Energy* (2014), and *Fracking* (2015). His other recent books include *Physics: Oryx Frontiers of Science Series* (2000); *Sick!* (4 volumes; 2000); *Science, Technology, and Society: The Impact of*

Science in the 19th Century (2 volumes; 2001); *Encyclopedia of Fire* (2002); *Molecular Nanotechnology: Oryx Frontiers of Science Series* (2002); *Encyclopedia of Water* (2003); *Encyclopedia of Air* (2004); *Nuclear Power* (2005); *Stem Cell Research* (2006); *The New Chemistry* (6 volumes; 2007); *Latinos in the Sciences, Math, and Professions* (2007); and *DNA Evidence and Forensic Science* (2008). He has also been an updating and consulting editor for a number of books and reference works, including *Chemical Compounds* (2005); *Chemical Elements* (2006); *Encyclopedia of Endangered Species* (2006); *World of Mathematics* (2006); *World of Chemistry* (2006); *World of Health* (2006); *UXL Encyclopedia of Science* (2007); *Alternative Medicine* (2008); *Grzimek's Animal Life Encyclopedia* (2009); *Community Health* (2009); *Genetic Medicine* (2009); *The Gale Encyclopedia of Medicine* (2010–2011); *The Gale Encyclopedia of Alternative Medicine* (2013); *Discoveries in Modern Science: Exploration, Invention, and Technology* (2013–2014); *Science in Context* (2013–2014); and *World of Physics* (2014).